Physics Problem Solving Secrets

120 Must-Know Problems Solved Step-By-Step

David J. Ulrich

To my beautiful wife and children

May you never lose your sense of wonder

Copyright © 2011 by David J. Ulrich

All rights reserved

Printed in the United States of America

ISBN 978-1-463-79865-9

PREFACE

If you are like most physics students, you are taking your class for some degree requirement. In other words, you are taking it because you have to. This doesn't mean that you are not interested in some sort of way. I've found the majority of students like to learn about these topics.

The issue is the actual doing of it. To be frank, learning enough physics to develop some level of competency is difficult. There is no substitute for long, diligent study. In particular: working the homework problems. One must learn how to work the problems in order to really *understand* this subject.

Now, that's no excuse for weak instruction or for a poorly formed curriculum. That's the gap this book hopes to fill. The focus of this book is solving problems. Not overly complicated ones—just examples of common exercises found in modern introductory physics classes. If you are looking for help, this is where to start.

The book presumes no calculus. So the prerequisite is some basic algebra. On the other hand, the book can still be useful for a calculus-based class since the physical reasoning is similar.

CONTENTS

1 Initial Advice — 1
   Where To Start 1
   Things You Figure Out After It's Too Late 5
   Dealing With The Math 11
   Calculus Fundamentals 12
   Trust Your Gut 14

2 Mechanics I — 17
   Free Fall 17
   Projectiles 23
   Inclined Plane 29
   Ropes And Pulleys 33
   Sliding Block With Friction 37
   Circular Motion 46
   Banked Curve 49
   Ideal Spring 52
   Pendulum 57
   Kepler's Third Law 59

3 Mechanics II — 61
   Simple Machines and Work Defined 61
   Horsepower 64
   Projectiles (Again) 65
   A Roller Coaster 69

Energy Lost To Drag	71
Turning Points	72
Collisions And Momentum	74
Ballistic Pendulum	79
Rocket Propulsion	81
Rain Pressure	83

4 Mechanics III · 85
- Tension And Torque · 86
- Double Support · 91
- Flywheel · 93
- Rolling Things · 98
- Stretching A Cable · 100
- Floating Things · 103
- Flowing Things · 106
- Tuning A Guitar · 109
- Intensity Of Sound · 111
- Wave Interference · 113

5 Heat and Light · 115
- Thermal Expansion · 115
- Calorimetry · 118
- Ideal Gas Law · 122
- Adiabatic Expansion · 124
- Carnot Heat Engine · 127
- What About Entropy? · 131
- Light On Glass · 135
- Lenses · 139
- Optical Systems · 141
- Wave Nature Of Light · 144

6 Electromagnetism · 151
- Field From Multiple Charges · 152
- Electric Potential · 158
- The Capacitor · 160
- Ohm's Law, Power Loss · 162
- Equivalent Resistance · 165
- Magnets, Motors, And Solenoids · 169
- Electric Generators · 180

	Transformers And Inductors	185
	AC Circuits	187
	Electromagnetic Waves	192
7	**Modern Physics**	**201**
	Improper Frames	202
	Twin Paradox	206
	Relativistic Fusion	210
	Photo-Electric Effect	216
	Spectroscopy	219
	De Broglie Waves	221
	Atomic Orbitals	224
	Nuclear Energy	227
	Radioactive Half-Life	232
	Quarks And The Standard Model	235
8	**Final Advice**	**241**
	Test Anxiety	241
	Stay Focused	243
Glossary		**245**
Index		**257**

Physics Problem Solving Secrets

CHAPTER

ONE

INITIAL ADVICE

You've been staring at the same homework problem for a quarter of an hour. It makes less sense than when you started. You need to step back and work through the fundamentals. This chapter will help by reminding you of them.

Where To Start

Most physics problems are word problems. So the first step is really just grammar. I mean more than simply finding a question mark—although that's a good start.

Job One: Find The Question

What are you being asked to deliver? An amount of energy, or a temperature, or an angle, or what? Realize that you don't have to *understand* a question to deconstruct it. Consider the following.

> **Problem**
>
> A 0.010 kilogram bullet is fired into a 1.0 kilogram block of wood. The wood is suspended from the ceiling by a thin rope and swings to a maximum height of 0.25 meters. What is the initial velocity of the bullet?

You are being asked for a "velocity".

> **Problem**
>
> Calculate the period of the Moon's orbit around the Earth. Assume it orbits with uniform circular motion and an acceleration equal to 0.00272 m/s^2.

Here the question is written in the form of a command. The thing you are being asked for is "the period".

Once the quantity of interest is identified, it is usually a matter of picking which equation you need in order to calculate it. Work backward from that quantity to the equation, to the quantities needed in the equation, etc. In this way you can usually work out how to answer the question.

Don't forget to pay attention to external clues. For example, what chapter is this question from? This can help limit the possibilities of where to start.

Dealing With Combination Questions

There are generally three levels of difficulty in physics homework. The easiest level is that in which you identify the relevant equation then plug and chug. There are a few of them here, but I have generally avoided this type of exercise.

Where To Start 3

The second level deals with actually *using* an equation in some context. Perhaps you have to calculate some intermediate data or you must use more than one equation to answer the problem. These are the type of problems dealt with in this book.

The third level is the most difficult. Typically these types of problems either require you to *derive* some equation or involve combining concepts from several different contexts. Since there are so many different possibilities, it's not really possible to cover all these combinations. This book will give you a firm fundamental approach that you use to at least begin a problem like this. Usually finding a way to begin is the hardest part.

Job Two: Find The Data

So you have an idea of what it will look like when the problem is done. You also have an idea how to begin. How do you fill in the middle? There are really two approaches. I think of it as working the problem backward and forward.

Usually, once you have the quantity of interest, the relevant equation will occur to you. That equation will require certain information. Look for this information in the statement of the problem. Keep an eye out for *implicit data* (see below). If it is there, you ready to go.

If not, you must calculate the required information from the data given. This means you must find a second or third equation to use. Look over your set of equations to find a match to your required quantities. This is what I call working the problem backward. You are moving from the desired quantity back into the starting data.

On other occasions, you will read the data and they will immediately suggest an equation to use. This is working the problem forward. This approach is often less successful since it is a little like walking in the dark. You take each step forward *hoping* to reach the answer.

Usually this direction is effective as a supplement to working the problem backward. In more difficult problems one might begin working backward then forward eventually meeting in the middle. If this happens to you, remember to rewrite the answer in a logical way to present a coherent argument.[1]

When There Is Not Enough Information

Although it can happen that you have too much information, it is really quite rare. More often than not, you have overlooked something. Remember you need one equation for each unknown. No more, no less.

But what if you don't have enough information? You never don't have enough information. Usually you are not seeing the *implicit data* in the problem. This is data hidden in the circumstances of the problem, but not explicitly stated. A classic case is in the following.

> **Problem**
>
> A bottle rocket is launched straight up. It reaches its maximum height in 4.0 seconds. How high does it go?

---

[1] I've often felt there is something slightly deceptive in presenting these solutions to students in such a buttoned-up, polished way. Textbooks and professors give the false impression that this is how the problems are solved. This couldn't be further from the truth. The reality is much more messy and laborious. But if the professor actually showed you *how* he arrived at the solution rather than the solution itself, the lecture would no doubt be quite confusing. Do not feel discouraged if your solution method is a mess. The issue is whether it is *correct*. Just remember to clean it up before you turn it in. Otherwise the grader won't be able to distinguish whether it is correct or not!

When a projectile is launched straight up, at the top of its trajectory it comes to a complete stop (for a moment). So, in this problem, the velocity is equal to zero at the maximum height. It is not stated explicitly in the problem, but follows from the way projectiles work.

Another common situation is that the data it appears you need actually cancels in the end. When something like this happens, it is superfluous for the author to include this data because it does not bear on the solution. This happens frequently with the mass of an object, for example.

If a quantity is missing that you apparently need, substitute it with a letter. This strategy is helpful in some combination problems, also. Sometimes you just need to push through to the end to see whether some quantity is relevant or not. And that is part of your learning experience too.

Things You Figure Out After It's Too Late

This section will help you avoid certain common pitfalls for beginners. Your professor may assume you are aware of certain basics without saying so.

Significant Digits

I have found that this subject is one of those that become more obscure the more you explain it. I used to spend a lot of time describing why 100.00 has five significant digits while 10000 only has one.

But then I realized something simple. There just aren't that many choices: the answer is usually either two or three. Rarely four, never one or five. So, really, the issue boils down to whether is it two, three, or four significant digits.

For some reason this makes the issue simpler. Look at the numbers in the statement of the problem. Do they have two, three, or four significant digits? Your answer should have the same number as the data with the fewest number of significant digits.

Now I know some of you will sweat over whether 100 has one, two, or three significant digits. If you don't get it, the truth is that it rarely makes a difference. This is because the other numbers in the problem will be enough to decide between two, three, or four.

Scientific Notation

There is no question about it. You have to get comfortable with scientific notation. Remember that a number in scientific notation consists of two parts: a coefficient (between one and ten) and an exponent. Any number can be written in scientific notation.

One thing that sometimes comes up is adding two numbers in scientific notation. For example,

$$3.4 \times 10^6 + 1.2 \times 10^4$$

This one you can actually do in your calculator. The way to figure it out on paper is to rewrite them with the same exponent.[2] In this case,

$$1.2 \times 10^4 = 0.012 \times 10^6$$

---

[2] This is something like finding common denominators when adding fractions. But easier—always use the larger exponent.

So
$$3.4 \times 10^6 + 0.012 \times 10^6 = 3.412 \times 10^6$$
If the exponents are very different, the sum becomes trivial. For example:
$$3.4 \times 10^{16} + 1.2 \times 10^4$$
But
$$1.2 \times 10^4 = 0.0000000000012 \times 10^{16}$$
So technically this sum is
$$3.4 \times 10^{16} + 0.0000000000012 \times 10^{16}$$
$$= 3.4000000000012 \times 10^{16}$$

Which is essentially the same as 3.4×10^{16}. This is why your calculator may not seem to add these two numbers. The second term is insignificant to the sum.

Inverse Trig Functions

There is a lot of trigonometry in physics. You will have to learn to use your sine, cosine, and tangent buttons. The difficulty sometimes is calculating the "inverse" functions.

When I ask you to calculate the tangent of 3.0°, no problem. Type **3.0** and hit **TAN**. But what if we have the reverse problem? Which angle has a sine equal to 0.5? This is what the inverse of the sine function is for.

There are two accepted ways to denote this calculation: $\arcsin(0.5)$ and $\sin^{-1}(0.5)$. The first is clearer, but the second is more common. The important thing to remember is that
$$\sin^{-1}(0.5) \neq \frac{1}{\sin(0.5)}$$

Degrees Versus Radians

Another trig-related issue is the unit used to measure angles. By far, the most common is the degree. Everyone knows there are 360° in one revolution. But, in physics, another common unit is the *radian*. By definition, there are 2π radians in one revolution.

The reason radians are often used is that the arc-length of one radian is equal to the radius of the circle. Since the radius of a circle is the natural unit of length for that circle, measuring the angle in these units simplifies many formulas. The most important example is

$$\sin x \approx x \quad (\text{small } x \text{ only})$$

This formula *only* works if the angle is measured in radians. For example, $\sin(0.1) = 0.0998$. (Try this with 0.1°.) This example is quite handy since, in effect, you get to ignore the trig if the angles are small.

Logarithms

Every once in a while you will need to use a logarithm. They show up directly when dealing with sound intensity (decibels) and in a few other places (like thermodynamics).

The logarithm is the inverse function to the exponential function. So, a logarithm answers the question, "To what exponent must I raise my base to get a particular number?" Thus, the base 10 logarithm of 100 is 2 is because $10^2 = 100$. There are two bases that get used so frequently they are on your calculator: base 10 and base $e = 2.71828...$ These are called the "common" and "natural" logarithm, respectively.

The most important thing to remember about logarithms is

that they turn multiplication into addition:

$$\log(ab) = \log(a) + \log(b)$$

The second most important thing to remember is that they turn exponents into multiplication:

$$\log(a^b) = b\log(a)$$

Whenever you use logarithms on paper, you will no doubt use these two facts.

The exponential function on your calculator (often labeled EXP) is the inverse of the natural logarithm. So it will calculate

$$\exp(x) = e^x$$

This shows that you can get the number e in your calculator by calculating $\exp(1)$. The exponential function comes up in contexts dealing with growth and decay. So, a question with "how long will it take ... to double" in it will usually involve a logarithm.[3]

Always Use SI Units

A unit is simply an arbitrary[4] standard chosen to convert physical qualities into mathematical quantities. When one measures something, it is by some comparison to the unit.[5] That comparison yields a number which we use in our equations, etc.

Since units are arbitrary, the system of units one chooses doesn't matter in the end. The physical relations are independent of the system of units. But in certain contexts certain

---

[3] The logarithm of 2, as a matter of fact.

[4] I don't mean chosen "willy-nilly", merely that it is a human convention. A lot of consideration goes into picking useful units.

[5] In a sense both relativity and quantum mechanics are a consequence of paying close attention to *how* things are measured.

units are easier to use than others. Usually this is done in order that some important constant is one, the speed of light, Planck's constant, etc.

Even though it doesn't matter, I recommend always using *SI units*. I have found that it is simply too easy to drop something if you are juggling your measurement units. I am willing to cheat occasionally with a nanometer here or a millisecond there, but never inches or pounds.

The nice thing about using a consistent system of units is that you can drop them from your calculations. Once every measurement is in the same system of units, the equations take care of the rest. Once you have completed the math, you simply put the relevant SI unit back. This is why I recommend, as the first step, to always convert your data into SI units.

Unfortunately, by dropping the units in the intermediate calculations, you do lose a double-check on your calculations: the ability to check your units in the end. If you keep the intermediate units you can manipulate them just like numbers. When that velocity you were looking for comes out in newtons per kelvin, you know you have done something wrong.

However, I have found this benefit is not as great as it sounds. Being able to forget about the units altogether is more valuable.

Converting Square and Cubic Units

One common stumbling block with units is converting square and cubic units. The proper way is the following.

$$(9.0 \text{ in}^2) \times \left(\frac{2.54 \text{ cm}}{1 \text{ in}}\right)^2 = 58 \text{ cm}^2$$

You *must* square the conversion factor! If I have a 1×1 square and I double the length of the sides, I double *both* the length and the width. I end up with four 1×1 squares in a grid.

Dealing With The Math

Same goes for cubic units.

$$9.0 \text{ cc} = (9.0 \text{ cm}^3) \times \left(\frac{1 \text{ m}}{100 \text{ cm}}\right)^3 = 0.00090 \text{ m}^3$$

Dealing With The Math

Physics is really applied math. The more you learn about physics, the more math you will need. In fact, it is a rare physics class that doesn't have a chapter or two developing some branch of mathematics. For introductory physics, it is usually a chapter on vectors.

Prerequisites

Here is a list of the math you should be familiar with going into your class. If you are not, make sure to have some sort of reference material handy.

- Algebra
  - Basic equation manipulation. Solving for one unknown.
  - Using the quadratic equation.
  - Solving two equations with two unknowns. This is somewhat rare, but it comes up once in a while.
- Geometry
  - Angles and triangles.
  - Trigonometry. Sine, cosine, and tangent at least.

- Solid geometry. For example: the volume of a sphere.

Depending on your class, you may be expected to know some statistics (for the labs mostly). Usually this is also covered in class quickly. The topics include:

- Mean, median, mode.
- Standard deviation, root-mean-square.
- Some simple error analysis.

Technically you don't need calculus for this book. But a simple understanding of calculus does help understand some of the concepts better. You can skip the next section if you like. Don't worry—you won't need to *do* anything with the calculus.

Calculus Fundamentals

Functions

Calculus is the mathematical study of change and dependence. When one quantity drives the value of another, we say that the second is a *function* of the first. The first is the *independent variable* and the second is the *dependent variable*. Two variables can be related in many ways, but when one drives the other, you have a function.

Differentials

Since physics is the study of how and why things change, calculus is the natural mathematical language for it. In every case, the laws of physics take the form, "When x changes, then

Calculus Fundamentals   13

y does too." This is summarized mathematically as

$$\Delta p = k \Delta q$$

Where p is the dependent, measured variable and q is the independent, control variable. When we change the value of q by one unit then p changes by k units.

Usually there is only a range of Δq in which this equation is valid. A common trick to extend this range is to allow the value of k change also. But there is always some small value of Δq under which this equation is valid for a constant k value. The physicist's challenge is to find that k. The experimentalist measures it, the theorist calculates it.

If we take the limit of this equation as Δq gets smaller and smaller, we begin to talk about *differentials*.

$$dp = k\, dq \Rightarrow \frac{dp}{dq} = k$$

The dp and dq are the differentials and their ratio is called the *derivative* (of p with respect to q). The derivative is a measure of how sensitive the dependence is between p and q.

Of course, differentials are not really numbers. If they were, they would have to be smaller than every positive number, but larger than zero—that's a no can do. By pretending that they are, we are using 18th century logic. Although less rigorous, this logic works fine when dealing with functions that are not pathological.[6]

Numerical Calculus

Numerical calculus takes these differentials and assumes they are a very small number. We can then "solve" for the motion

---

[6] This is a real mathematical term.

of the dependent variable by iterating forward the independent variables by 0.001 units or whatever.

This is simple is principle, but small errors magnify quickly when taking this approach. A whole branch of mathematics is devoted to optimizing this iterative processing for speed and accuracy.

The nice thing about these numerical calculations is that they can be easily modeled in a spreadsheet program. This gives even non-programmers a way to solve problems that are unsolvable even for calculus students.

On the other hand, this yields the solution for only one particular example of one particular problem. Change a single number and the calculation must be redone. Having the physics summarized in closed mathematical form allows one to *understand* the relationships in a way that is difficult with numerical calculus.

Trust Your Gut

Finally, if I had one sentence to give in terms of advice, it would be to "trust your gut". You deal with physics everyday. It's ingrained in your subconscious. Every time you lift a pencil or lift a suitcase, you are performing some kind of physical calculation in your mind. Release that intuition.

Also, learn to reason by order of magnitude. If the problem starts talking about centimeters and kilograms, be sure to recognize that these are normal, everyday units. When you get to the end and the answer is 1.5×10^8 m/s, know that this is probably wrong (it's half the speed of light). One meter per second is quick: about the speed of the flick of your hand. But this answer is eight orders of magnitude larger. I know it's probably wrong, and I don't even know the question!

Trust Your Gut 15

Picture in your mind these SI units. How heavy is a kilogram? (An apple.) How large is a cubic meter? (A dishwasher.) This will develop your intuition and help you avoid egregious errors.

CHAPTER

TWO

MECHANICS I

Free Fall

Motion happens in the three dimensions of space. So the motion of an object is specified by both its speed and direction. *Velocity* is the term used to specify both. By definition, *acceleration* is the rate at which velocity changes.

This means that when we are told that an object is accelerating either the direction or the speed is changing. In the real world it is common for both to happen.[1] In the textbook, the acceleration often only changes the speed. The first five problems in this chapter focus on this situation.

By definition, the average velocity is

$$\bar{v} = (\Delta x)/(\Delta t) \qquad (2.1)$$

---

[1] If the acceleration is not constant, most of the time you need calculus to solve the problems and even then only for very special cases. Chapter 3 introduces an alternate solution method (using energy) for more general cases.

and the average acceleration is

$$\bar{a} = (\Delta v)/(\Delta t) \tag{2.2}$$

When the acceleration is constant, we have $a = \bar{a}$. We can rewrite Equation (2.2) in three ways:

$$\begin{aligned} v_f &= v_i + a(\Delta t) \\ v_i &= v_f - a(\Delta t) \\ v_f - v_i &= a(\Delta t) \end{aligned} \tag{2.3}$$

In addition, the velocity increases at a *constant* rate, so the average velocity is given by

$$\bar{v} = \tfrac{1}{2}(v_f + v_i) \tag{2.4}$$

Using the definition of average velocity, Equation (2.1), we can write:

$$\Delta x = \tfrac{1}{2}(v_f + v_i)(\Delta t) \tag{2.5}$$

It is more common to represent Δt as t and Δx as x. You can either justify this as mere shorthand, or recognize that this implicitly sets $t_i = 0$ and $x_i = 0$. Either way, these equations represent the starting point for dealing with problems of constant acceleration.

We can combine these equations to yield three more:

$$x = v_i t + \tfrac{1}{2}at^2 \tag{2.6}$$

$$x = v_f t - \tfrac{1}{2}at^2 \tag{2.7}$$

$$v_f^2 - v_i^2 = 2ax \tag{2.8}$$

Equations (2.3), and (2.5)–(2.8) are all you will need to solve *any* problem dealing with constant acceleration. A key step

Free Fall

in solving these problems is to identify which equation to use. Fortunately it's pretty easy. I'll show you how.

Free-fall is a special case of constant acceleration. Since the force of gravity is a constant (over "short" distances like a kilometer), problems dealing with objects falling under the acceleration due to gravity use the previous formulas. On earth, this "g-force" is -9.8 m/s^2. (The negative sign indicates the acceleration is *down*.) On the moon this "g-force" is -1.6 m/s^2.

Our first example problem is:

Free-Fall From Rest – How Far?

A metal sphere, initially at rest, falls for 2.0 seconds. How far does it drop?

Answer: -20 meters

The question is "how far" ... i.e., x. (It should be negative when we are done, right?). We are also told the initial velocity ("initially at rest") and the duration ("2.0 seconds"). In summary, we have:

$$x = ??$$
$$v_i = 0$$
$$t = 2.0$$
$$a = -9.8$$

Notice how these four variables are in Equation (2.6). We can use this equation to solve for x:

$$x = (0) + (0)(2) + \tfrac{1}{2}(-9.8)(2)^2 \Rightarrow x = -19.6$$

Since there are two significant digits in the problem, the answer is rounded to -20 meters.

Every other problem can be solved in the *same* way:

- Which unknown variable are we looking for: x, v_i, v_f, a, t?

- What other three variables do I have information about?

- Which of the five equations involve these four variables?

- Solve this equation for the unknown (always use sig figs!)

Let's look at a similar problem:

Free-Fall From Rest – How Long?

A coin is dropped into a well and drops 10 meters. How long does it fall before hitting the ground?

Answer: 1.4 seconds

The data in the problem are:

$$x = -10$$
$$v_i = 0$$
$$t = ??$$
$$a = -9.8$$

Notice how I had to read between the lines to get $v_i = 0$. This is an example of *implicit data*. Again, we can use Equation (2.5).

$$(-10) = (0) + (0)(t) + \tfrac{1}{2}(-9.8)(t)^2$$
$$\Rightarrow t = \sqrt{10/4.9} = 1.4286$$

So much for the easy problems. Let's look at another:

Free Fall

Free-Fall – How Fast From Duration?

A bottle rocket is launched straight up. It reaches its maximum height in 4.0 seconds. How high does it go? How fast was it launched?

Answer: (a) 78 meters (b) 39 m/s

The data in the problem are:

$$x = ??$$
$$v_f = 0$$
$$t = 4.0$$
$$a = -9.8$$

I have picked this problem to show you another example using *implicit data*. The key is the phrase "maximum height". At the top of the trajectory the vertical motion (for a brief moment) stops. In symbols: $v_f = 0$.

The equation to use is Equation (2.7):

$$x = (0)(4.0) - \tfrac{1}{2}(-9.8)(4.0)^2 \Rightarrow x = 78.400$$

Now for the second part. What do we know?

$$v_i = ??$$
$$v_f = 0$$
$$t = 4.0$$
$$a = -9.8$$

We can use Equation (2.3)...

$$0 = (v_i) + (-9.8)(4.0) \Rightarrow v_i = 39.200$$

Occasionally a problem will come up that does not involve time explicitly:

> **Free-Fall – How Fast From Distance?**
>
> A bottle rocket is launched straight up. It reaches a maximum height of 20 meters. How fast was it launched?
>
> **Answer**: 20 m/s

The data in the problem are:
$$v_i = ??$$
$$x = 25$$
$$v_f = 0$$
$$a = -9.8$$

We need to use Equation (2.8).
$$(0)^2 - (v_i)^2 = 2(-9.8)(20) \Rightarrow v_i = \sqrt{392} = 19.799$$

Let's finish with a more complicated problem...

> **Free Fall On The Moon**
>
> Let's suppose when a person jumps, he is in the air for one second. All else equal, how long would the same person be airborne on the moon?
>
> **Answer**: Six seconds

Find the question: "how long". What do we know?
$$a = -1.6$$
$$t = ??$$
$$x = 0$$
$$v_i = \text{TBD}$$

Two lessons from this list. First we have another piece of *implicit data*. The jump starts and stops at the same point, so

Projectiles

the net displacement is zero. Though not in this problem, a keyword to look for is "range". It implies the beginning and ending height is the same.

The second lesson is that this is a two part question in disguise. We were able to extract a, t, and x directly from the problem. But we need one more quantity. We also know we have not used all the data from the problem. The "all else equal" seems to indicate that the initial velocities are the same on the moon and the earth. Let's see if we can extract this information from the data concerning the earth jump.

$$a = -9.8$$
$$t = 1$$
$$x = 0$$
$$v_i = ??$$

Now we have a standard problem. We use Equation (2.3) to yield:

$$(0) = (v_i)(1) + \tfrac{1}{2}(-9.8)(1)^2 \Rightarrow v_i = 4.9000$$

We can take this for the moon jump using Equation (2.3):

$$(0) = (4.9)(t) + \tfrac{1}{2}(-1.6)(t)^2 \Rightarrow t = 6.1250$$

Your hang-time on the moon would be over six seconds!

Projectiles

We spent quite a bit of time on the previous type of problem because it will be used again and again. Often the endgame for other problems will have the same form as these free-fall problems. Like projectiles.

Projectile problems also use the free-fall solution. Charting out the path of a projectile is an ancient problem. Literally.

You can find a discussion of it in Aristotle. In more modern times, Galileo studied the problem and recognized that it can be analyzed using the previous techniques for free-fall.

The trick is to decompose the problem into two pieces: horizontal motion and vertical motion. There is no force of gravity in the horizontal dimension, so the horizontal velocity does not change (zero acceleration). For the vertical dimension we use the same techniques as before. These motions occur at the same time.

> **Projectile Range**
>
> A projectile is launched at a 60° angle at 20 m/s. How far does it travel?
>
> **Answer**: 18 meters

The first step is to break the motion into its two parts. Here is where we need some trig (usually a diagram helps here). The components of the initial launch are

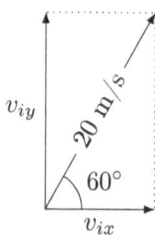

Figure 2.1: Components of initial velocity

$$v_{ix} = v_i \cos\theta = (20)(\cos 60°) = 10.000$$
$$v_{iy} = v_i \sin\theta = (20)(\sin 60°) = 17.321$$

The horizontal motion is simpler because there is no acceleration. What do we know? If we cast this in a way that is similar

Projectiles

to the previous problems, we have:

$$v_{ix} = 10.0$$
$$a = 0$$
$$v_{fx} = 10.0$$
$$x = ??$$

It looks like we ought to be able solve this directly. But no, this is a special case. The formula we look to is Equation (2.8) but all we end up with is $0 = 0$ because the acceleration is zero. Our only option is to calculate t and use Equation (2.1). But how do we get it? We turn to the vertical.

$$v_{iy} = 17.3$$
$$a = -9.8$$
$$y = 0$$
$$t = ??$$

Now, this is a problem we can solve. We use Equation (2.6).

$$(0) = (17.3)(t) + (-9.8)(t)^2 \Rightarrow t = 1.7653$$

We are now in a position to use Equation (2.1):

$$x = (10.0)(1.7653) = 17.653$$

The range of the projectile is just under 18 meters.

The following is an example of a more difficult projectile problem. I'll use it to explain some difficulties you may encounter in other problems.

Projectile Angle

A target is 100 meters away on level ground. Suppose you have a cannon that can fire a cannonball at 50 m/s. At what angle should you aim the cannon to strike the target?

Answer: Either 12° or 78°

The angle of the cannon is the angle of the initial velocity. If we knew either component of the velocity we would be done since we could use $v_{ix} = v_i \cos\theta$ or $v_{iy} = v_i \sin\theta$ to solve for the angle θ.

What about the horizontal direction? From the previous solution we know that we need the time t in order to get anywhere, so let's begin with the vertical...

$$v_{iy} = (50)(\sin\theta)$$
$$a = -9.8$$
$$y = 0$$
$$t = ??$$

You can see the dilemma. We don't actually know v_{iy}. This is the first point that makes this problem difficult. Occasionally, it is not possible to solve a problem directly like we have done so far. Notice that in all the previous solutions the next-to-last step is an equation involving the quantity in which we are interested. If the problem is "easy", this equation is one of the fundamental ones (like $x = v_i t + \frac{1}{2}at^2$).

In other words, the key step has been to *identify* the equation to use. In this problem we need to *create* the equation to use. In order to do this, we ignore the fact that we don't know θ and push forward as if we do.

The equation to use here is Equation (2.6).

$$(0) = (50)(\sin\theta)(t) + \tfrac{1}{2}(-9.8)(t)^2$$

This is as far as we can go. We need to find another equation using t and substitute it into this equation. The information we have left is about the horizontal direction. Since there is no acceleration, the only equation available is Equation (2.1).

$$(100) = (50)(\cos\theta)(t)$$

At this point we have two equations with two unknowns. This is good. Mathematically the problem is solvable. We have

Projectiles

extracted enough physical information from the problem to solve it. We can now ignore the physics and hunker down with the algebra.

The way to solve the math problem of two equations with two unknowns is:

- Solve one of the equations for one of the variables. Since we can choose, choose the one you are *least* interested in.

- Substitute the variable in the other equation.

- Solve for the variable.

- Substitute the answer into the first equation.

- Solve for the other variable—the one you are *most* interested in.

In our case, we solve the second equation for t.

$$(100) = (50)(\cos\theta)(t) \Rightarrow t = \frac{2}{\cos\theta}$$

And substitute...

$$(0) = (50)(\sin\theta)(t) + \tfrac{1}{2}(-9.8)(t)^2$$

$$\Rightarrow (0) = (50)(\sin\theta)\left(\frac{2}{\cos\theta}\right) + \tfrac{1}{2}(-9.8)\left(\frac{2}{\cos\theta}\right)^2$$

But we don't need to finish the math (i.e., solve for t) to answer our question. We now have a single equation involving θ. This was our goal: to *create* an equation to use. So we are on the right track.

The problem we now face is that these trig functions are in our way. Generally, solving equations with trig functions can be difficult simply because there are so many options and tricks available. The tricks most often used are (in order of popular-

ity):
$$\cos^2\theta + \sin^2\theta = 1 \tag{2.9}$$

Divide this equation by $\sin^2\theta$ or $\cos^2\theta$ for other versions. The second type of trig identity is the double angle formulas.

$$\begin{aligned}\sin 2\theta &= 2\sin\theta\cos\theta \\ \cos 2\theta &= 2\cos^2\theta - 1\end{aligned} \tag{2.10}$$

Finally, we have the following half-angle formulas.

$$\begin{aligned}\sin(\theta/2) &= \sqrt{(1-\cos\theta)/2} \\ \cos(\theta/2) &= \sqrt{(1+\cos\theta)/2}\end{aligned} \tag{2.11}$$

Back to our problem. Our first step is to get that pesky $\cos\theta$ out of the denominator by multiplying by $\cos^2\theta$.

$$(0) = (50)(\sin\theta)\left(\frac{2}{\cos\theta}\right) + \tfrac{1}{2}(-9.8)\left(\frac{2}{\cos\theta}\right)^2$$
$$\Rightarrow (0) = (100)(\sin\theta)(\cos\theta) - 19.6$$

Notice the $(\sin\theta)(\cos\theta)$. That brings to mind the double-angle formula for $\sin 2\theta$. Let's manipulate the equation to prepare for this substitution.

$$\begin{aligned}(0) &= (100)(\sin\theta)(\cos\theta) - 19.6 \\ \Rightarrow (0) &= (50)(2\sin\theta\cos\theta) - 19.6 \\ \Rightarrow (0) &= (50)(\sin 2\theta) - 19.6\end{aligned}$$

We can now solve for $\sin 2\theta$ then θ.

$$\begin{aligned}(0) &= (50)(\sin 2\theta) - 19.6 \\ \Rightarrow \sin 2\theta &= 0.392 \\ \Rightarrow \theta &= \tfrac{1}{2}\sin^{-1}0.392\end{aligned}$$

These inverse trig functions can be tricky. Remember they answer the question: what angle has a sine of 0.392? The

Inclined Plane

answer is 23.1° (this is probably the one you get with your calculator), but 156.9° also works (this is 180° − 23.1°). This means there are two answers[2] to this problem:

$$\theta = \tfrac{1}{2}\sin^{-1} 0.392 \Rightarrow \theta = \tfrac{1}{2}(23.1° \text{ or } 156.9°)$$

In general, the formula for the range of a projectile is:

$$R = (v^2/g)\sin 2\theta \qquad (2.12)$$

Inclined Plane

The two previous sections are usually classified as problems in *kinematics*, the study of motion. Now we begin our study of the *causes* of motion. This goes by the name *dynamics*. First, we will study the nature of force. Force is directional, a vector. An inclined plane is a good example of the vector nature of force.

When we combine multiple forces, the direction needs to be taken into account. Once the net force is determined, Newton's Second Law states that the acceleration of the particle obeys the following formula:

$$F_{\text{net}} = ma \qquad (2.13)$$

The larger the mass, the smaller the resulting acceleration. If we know the resulting acceleration we can deduce the size of the force causing it by multiplying by the mass. This is why the force of gravity (otherwise known as weight) is given by the formula

$$W = mg \qquad (2.14)$$

---

[2] By the way, suppose you didn't know about the double-angle formula above? One can always solve these problems by using the first trig identity to substitute out $\cos\theta$ using $\sqrt{1-\sin^2\theta}$. You will find that you end up with a quadratic formula in $\sin\theta$ (after much pain and suffering), but you will get the same two answers in the end.

There are four forces involved in these basic force problems.

- Gravity: Always down.

- Support: Always pushes.

- Tension: Always pulls.

- Friction: Always resists.

All of these forces can be considered constant. This allows us to apply the solution methods of the previous problems to situations involving these forces.

Consider a block on a frictionless inclined plane. There are two forces in play. The force of gravity and the support force. If the plane were not there, the block would fall. But the plane is there and counterbalances the force of gravity. The critical thing is that this support force *only* points perpendicular to the surface.[3] So the two forces are not aligned. The support force balances the component of gravity perpendicular to the surface. The component parallel to the surface remains and accelerates the block.

Remember that the force due to gravity is the mass of the block times the acceleration due to gravity (9.8 m/s^2).[4]

The support force is variable. There is no simple formula for it—it changes for different situations. The support is the force required to counter-balance the force *into* the surface. If the force into the surface is large, the support force is large. If the force into the surface is small, the support force is small. The

---

[3] This is true in general for support forces. This is why they are sometimes called "normal" forces. (Normal being a term in mathematics for perpendicular).

[4] We have dropped the sign here because we are going to keep track of the direction of these forces in a different way than in the previous sections.

Inclined Plane

support force must be calculated based on the circumstances in the problem.

Let's show these principles in action.

> **Inclined Plane – No Friction**
>
> A block is sliding without friction on a plane inclined at 30° sloping up to the right. What is the acceleration of the block?
>
> **Answer**: -4.9 m/s^2

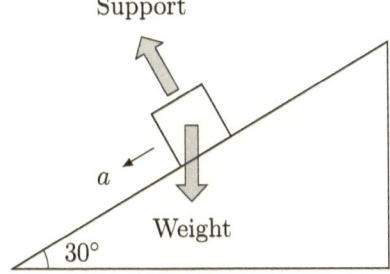

Figure 2.2: Frictionless inclined plane

Again, the two forces here are gravity and the support force. However, we are not given the mass of the block. It is common for a solution to be independent of the mass, so don't be surprised by this. Just label the mass m and keep it in the algebra. It should cancel out in the end.

The most important step here is the first one. We *redefine* the horizontal x-direction as the direction along the slope of the plane. When we do this, the support force is entirely in the y-direction. There is no support force in the x-direction.

Now, the support force balances the component of the weight

perpendicular to the plane, but we don't need to calculate it. We know the support and weight component cancel out, so they can't affect the motion. This is why we orient the x and y-directions this way.

We need to calculate the component of the weight in the x-direction (parallel to the plane).

$$W_x = W \cos\theta = (mg)(\cos\theta) = (m)(9.8)(\cos\theta)$$

But what is θ? My advice is to *always* measure your angles from the positive x-direction (whatever that is). If you do so, the equation to find components is *always* given by

$$A_x = A \cos\theta$$
$$A_y = A \sin\theta$$

In our case, we start our angle measurement "30° sloping up to the right", circle up and around to the straight-down (the direction of the weight).

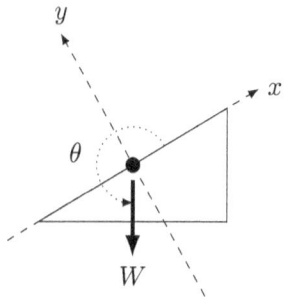

Figure 2.3: How to measure your angles...

This is a total of 240°. Therefore,

$$\begin{aligned} W_x &= (m)(9.8)(\cos 240°) \\ &= (m)(9.8)(-0.5) \\ &= (-4.900)(m) \end{aligned}$$

Notice how the sign is correct: the force is to the *left* with a magnitude of 4.9 newtons. The signs work out naturally if you follow the advice I gave about angle measurement.

According to Newton's Second Law, the acceleration is F/m. So the net acceleration is

$$F/m = (-4.900)(m)/(m) = -4.900$$

The mass cancels out, just as advertised. If the block had an initial velocity up the slope, it would slow down, reach a maximum height, and begin to slide down to the left, just like the free-fall problems earlier.

Ropes And Pulleys

> **Block and Tackle – Trick Question**
>
> A block with a mass of 10 kilograms hangs from a rope and pulley attached to the roof. The block rises with a constant velocity of 2 m/s. What is the tension on the rope?
>
> **Answer**: 98 newtons

This is a trick question. This is an example of a *moving* system in equilibrium. Equilibrium means that all the forces balance. If there is no net force, there is no acceleration. Therefore, the velocity is constant. So in this problem, there is no net force. The tension must counter-balance the weight. Thus,

$$T = W = mg = (10)(9.8) = 98$$

Block and Tackle – Equilibrium

Suppose a pulley is attached to a 100 kilogram mass on the ground. A rope attached to the ceiling runs around this pulley (#1) and through another pulley (#2) also attached to the ceiling. How much tension is required to lift the mass?

Answer: 490 newtons

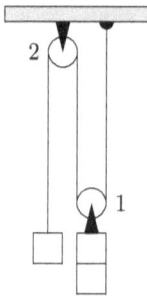

Figure 2.4: Simple block and tackle with mechanical advantage of two

Three items. First, it may not look like it, but this is an equilibrium problem. Equilibrium occurs right before the maximum tension when the mass lifts. So we are looking for a situation in which the net force is zero. Any tension greater than that equilibrium will lift the mass.

Second, an ideal rope has the following property: no matter what the orientation, no matter where along the rope, as long as the rope is taut, the tension is the same. Technically, tension is a measure of the forces pulling apart an object at a particular point. But for an ideal rope, the tension is the same everywhere, so this distinction is sometimes overlooked. Thus, the tension in the rope from the ceiling to pulley #1, and the tension between pulleys, and the tension after pulley #2 are all the same value.

Ropes And Pulleys

Third, the pulley attached to the block (#1) has *two* ropes pulling up. This means that the tension in the rope need only be *half* the weight to lift the block. This is the answer:

$$2T = mg \Rightarrow T = (100)(9.8)/2 = 490$$

Because of the way the pulleys are combined only half the force is required to lift the weight. On the other hand, when one pulls the rope the weight only rises half the height. In physics, *work* is the product of force times distance. In this case, the work done lifting the mass against gravity equals the work done by creating tension in the rope. This will come up again in the next chapter.

One final pulley exercise.

Atwood Machine

Suppose two masses that are connected by a rope hang over a pulley. The first mass is 10 kilograms, the second mass is 5.0 kilograms. At what rate does the 10 kilogram mass fall?

Answer: 3.3 m/s^2

This is a classic problem called the *Atwood machine*.

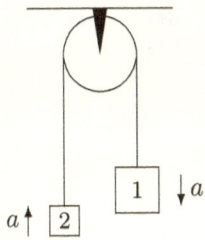

Figure 2.5: Atwood machine

You can imagine what happens: the larger mass falls, the smaller mass rises. They both accelerate until the floor or ceiling is hit. But where to begin when solving this problem?

We know that the tension is the same throughout the rope. But the tension pulling up the larger mass is less that its weight because it will fall under the influence of gravity. Similarly, the tension pulling up the smaller mass is greater than its weight because it will rise in spite of gravity. That puts some boundaries on the solution but doesn't really get us much farther.

It is easy to overlook the other *implicit data* which is typical in these rope and pulley problems: an ideal rope does not change its length. This meadns that if the smaller mass rises with an acceleration of a, the larger mass must fall with the same magnitude of acceleration, $-a$.[5] Using Newton's second law on each mass we can write:

$$F_{\text{net}} = T - m_1 g = m(-a) \Rightarrow T - (10)(9.8) = (10)(-a)$$
$$F_{\text{net}} = T - m_2 g = m(a) \Rightarrow T - (5)(9.8) = (5)(a)$$

We have two equations, two unknowns. Solve the first for T:

$$T - (10)(9.8) = (10)(-a) \Rightarrow T = -10a + 98$$

Then substitute it into the second equation and solve for a:

$$T - (5)(9.8) = (5)(a)$$
$$\Rightarrow (-10a + 98) - 49 = 5a$$
$$\Rightarrow 49 = 15a$$
$$\Rightarrow a = 49/15 = 3.267$$

In general, the solution is

$$a = g\left(\frac{m_1 - m_2}{m_1 + m_2}\right)$$

---

[5] Negative because it is falling *down*.

Sliding Block With Friction

Until now we have ignored friction. Everything has been "ideal". But friction is complicated. So, we introduce an idealized form of friction![6]

The idea is that friction is a consequence of two surfaces rubbing against one another. The more the surfaces are compressed, the larger the friction. In an equation we have:

$$F = \mu N$$

where N is the force pressing the surfaces together. In general, this is the same as the support force (or "normal" force) we have mentioned earlier. The symbol μ is simply a number: it must be calculated from experiment. It depends on a variety of circumstances, but primarily the nature of the two surfaces (are they glass, wood, wood and glass, etc.) and whether the object is moving or not. If the object is not moving, this is called *static friction*. If the object is moving, this is called *kinetic friction*.

Frequently, in situations where the dynamics are complex, physicists dramatically simplify the interactions using some sort of proportionality like this.[7] Really these equations say something like: "We know that when x goes up y does too. In fact when x goes up one unit, y goes up k units. Let's summarize this as $y = kx$." Often this is the first step in scientific inquiry. The next step is to explain why k has a particular value based on more fundamental considerations.

Back to problem solving. Let's look at an example with fric-

---

[6] We won't go beyond this. We could talk about air drag and other forms of friction, but the typical introductory class has a lot of fish to fry, so these topics are generally dropped.

[7] You will see it again when we talk about heat and current flow, for example.

tion.

> **Inclined Plane – With Friction**
>
> A block is on a plane inclined at 30° sloping up to the right. The coefficient of static friction is $\mu_s = 0.30$. Does the block slide or not?
>
> **Answer**: Yes, the block slides

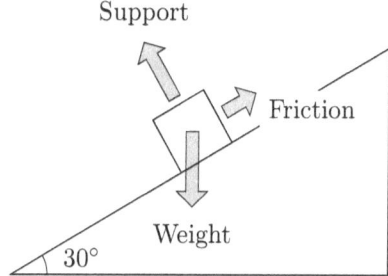

Figure 2.6: Inclined plane with friction

This is the same situation as the previous inclined plane problem but now we introduce friction. In that case we did not need to calculate the support force. Now we need it in order to calculate the friction involved. The support force counter-balances the y-component of the weight.

$$W_y = (mg)(\sin\theta) = (m)(9.8)(\sin 240°)$$
$$= (m)(9.8)(-0.866) = (-8.487)(m)$$

The negative sign shows that the weight is pointing *into* the inclined plane. The support force counter-balances this force:

$$N = (8.487)(m)$$

The maximum static friction force sustainable by the block is:

$$F_s = \mu_s N = (0.30)(8.487)(m) = (2.5461)(m)$$

Sliding Block With Friction

Is this sufficient to overcome the component of weight pulling the block down the plane? Looking back, that force was $(-4.9)(m)$ newtons of force. The static friction force is insufficient, so the block will slide.

What if μ_s were 0.60? Then the maximum static frictional force would be $F_s = (5.0922)(m)$. This would be sufficient to counter-balance the $(-4.9)(m)$ of residual weight. However, if this were to happen, the static friction force would not be $(5.0922)(m)$. If it were, the friction would push the block *up* the plane! The friction force is simply that which is necessary to balance the other forces (much like the support forces). That is, F_s would be $(4.9)(m)$.

Inclined Plane – Sliding With Friction

A block is on a plane inclined at 30° sloping up to the right. The coefficient of kinetic friction is $\mu_k = 0.30$. At what rate does it accelerate?

Answer: 2.4 m/s^2

In this case all the calculations are the same, so $F_k = (2.5461)(m)$. But kinetic friction is different than static friction. It doesn't "counter-balance"—it merely resists. This means that the value is what it is, but the sign opposes the other forces. In this case, since the component of weight is negative, the friction is positive...

$$F_{\text{net}} = W_y - F_s = (-4.9)(m) + (2.5461)(m)$$
$$= (2.3539)(m)$$

And according to Newton's Second Law,

$$a = F_{\text{net}}/m = (2.3539)(m)/(m) = 2.3539$$

Now, we combine everything we've learned so far in one wicked-long problem.

Inclined Plane With Pulley

Two masses are attached with a string hanging over a pulley. The pulley sits on the upper corner of a plane inclined at 20° to the right. The first mass is 1.0 kilogram and sits on the plane with $\mu_s = 0.50$ and $\mu_k = 0.30$. The second mass is 2.0 kilograms and hangs over the edge. Initially everything is at rest. Describe the situation one second later.

Answer: $a = 6.34$ m/s^2, $v = 6.34$ m/s, and $\Delta x = 3.17$ meters

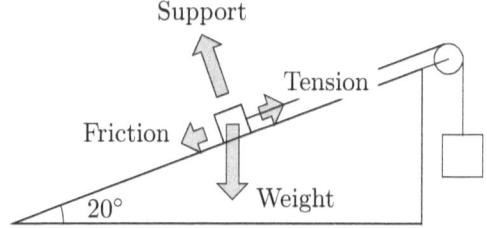

Figure 2.7: Inclined plane with pulley

In complex problems like this it is important to follow an algorithm lest one get lost in the details.

- Determine the forces acting on each important point in the system (usually this involves a free-body diagram).

- Calculate the components of each force along a "properly" oriented system.

- Add the components to find the net force acting on each of the points.

- Apply Newton's Second Law. If you know the acceleration of the bodies, plug them in. If you don't, you are going to have to solve for them.

Sliding Block With Friction

- Verify that there are enough equations for all of the unknowns. If not, look for *implicit data*.

- Solve the equations for the acceleration of all the bodies.

- Apply the free-fall principles to answer the specific questions in the problem.

An additional complication in this problem is that it does not tell us whether the system is in motion or not. If the system is in motion, we use kinetic friction. If not, we use static friction. When it is unclear like this, first assume the system is in equilibrium. This means that we assume the presence of a static friction force:

$$F = \alpha \mu_s N$$

where α is a number between 0 and 1. In the end, we will need to solve for this number α. If it is in the range between 0 and 1 we are fine. If not, the forces in the system are strong enough to overcome the maximum static friction force so the system is in motion. We then need to *rework* the problem with

$$F = \mu_k N$$

Fortunately, most of the calculations are identical to the static situation, so that is not as bad as it sounds. But we must always eliminate the static solution first before assuming the system is in motion.

For our problem, we have two masses. I'll show you the procedure in excruciating detail on the simpler one so you can see the process. (So sit back and relax: this is the longest solution in the book.)

The forces on the hanging mass are weight and tension. The simplest coordinate system is the regular one: x horizontal, y

vertical. For the weight we have:[8]
$$W = mg = (2.0)(9.8) = 19.6$$
The weight is straight down and the tension is straight up, so there is no horizontal component to worry about. Since we are assuming the system is at rest, we have
$$T - 19.6 = 0$$
This tells us that the tension in the rope is equal to the weight of the mass. This true because the block is at rest. Now for the sliding mass.

The forces on this mass are weight, support, tension, and friction. The simplest coordinate system is aligned with the plane: x parallel, y perpendicular. For its weight we have:

$\vec{W} = mg$ at $250°$
$\Rightarrow W_x = (m_1 g)(\cos 250°) = (1.0)(9.8)(-0.3420) = -3.352$
$\Rightarrow W_y = (m_1 g)(\sin 250°) = (1.0)(9.8)(-0.9397) = -9.209$

For its support we have:
$$\vec{N} = N \text{ at } 90°$$
$$\Rightarrow N_x = (N)(\cos 90°) = (N)(0) = 0$$
$$\Rightarrow N_y = (N)(\sin 90°) = (N)(1) = N$$

For its tension we have:
$$\vec{T} = T \text{ at } 0°$$
$$\Rightarrow T_x = (T)(\cos 0°) = (N)(1) = T$$
$$\Rightarrow T_y = (T)(\sin 0°) = (N)(0) = 0$$

For its friction we have:

$\vec{F} = \alpha \mu_s N$ at $0°$
$\Rightarrow F_x = (\alpha \mu_s N)(\cos 0°) = (\alpha)(0.5)(N)(1) = (\alpha)(0.5)(N)$
$\Rightarrow F_y = (\alpha \mu_s N)(\sin 0°) = (\alpha)(0.5)(N)(0) = 0$

---

[8] Since the weight is down, we include it with a negative sign.

Sliding Block With Friction

How do we know that the friction points *up* the plane? We don't. But if after solving for α we get a negative answer, we will know the friction actually points in the opposite direction. So we simply choose then wait and see.

The components of the net force on the sliding mass are:

$$\begin{aligned}F_{\text{net},x} &= W_x + N_x + T_x + F_x \\ &= (-3.352) + (0) + (T) + (\alpha)(0.5)(N) \\ &= -3.352 + T + (\alpha)(0.5)(N)\end{aligned}$$

and

$$\begin{aligned}F_{\text{net},y} &= W_y + N_y + T_y + F_y \\ &= (-9.209) + (N) + (0) + (0) \\ &= -9.209 + N\end{aligned}$$

Next we apply Newton's Second Law to the system. Remember, we are assuming a static solution, so $a = 0$.

$$F_{\text{net}} = ma = 0$$

After substitution, we have:

$$\begin{aligned}F_{\text{net},x} &= -3.352 + T + (\alpha)(0.5)(N) = 0 \\ F_{\text{net},y} &= -9.209 + N = 0\end{aligned}$$

We now have our formulas for each block. One from the hanging mass and two from the sliding mass. Collecting these, we have

$$\begin{aligned}&T - 19.6 = 0 \\ &-3.352 + T + (\alpha)(0.5)(N) = 0 \\ &-9.209 + N = 0\end{aligned}$$

We have three equations and three unknowns so this is solvable. The first and last yield:

$$\begin{aligned}T - 19.6 = 0 &\Rightarrow T = 19.6 \\ -9.209 + N = 0 &\Rightarrow N = 9.209\end{aligned}$$

Plugging these into the second equation gives us:

$$-3.352 + (19.6) + (\alpha)(0.5)(9.209) = 0$$

We can now solve for α.

$$(\alpha)(0.5)(9.209) = -16.248 \Rightarrow \alpha = 3.53$$

Since α is more than one, the maximum static friction force is insufficient to maintain equilibrium. This means we move on to phase two. (Are you surprised?) Assume the system is accelerating and that the kinetic friction force is in play.

Again, the forces on the hanging mass are weight and tension. But now when we apply Newton's Second Law, there is an acceleration of some sort. It's probably down. (If not, in the end, the final sign on the acceleration will correct us.) Thus,

$$T - 19.6 = (2.0)(-a)$$

The acceleration is negative because the motion is *down*.

The forces on the sliding mass are still weight, support, tension, and friction. The only difference here is in the friction force. We use the formula $F = \mu_k N$. Since we are assuming the motion is to the left, this friction must point to the right (same as before).

For the friction we have:

$$\vec{F} = \mu_k N \text{ at } 0°$$
$$\Rightarrow F_x = (\mu_k N)(\cos 0°) = (0.3)(N)(1) = (0.3)(N)$$
$$\Rightarrow F_y = (\mu_k N)(\sin 0°) = (0.3)(N)(0) = 0$$

The components of the net force on this mass are:

$$\begin{aligned} F_{\text{net},x} &= W_x + N_x + T_x + F_x \\ &= (-3.352) + (0) + (T) + (0.3)(N) \\ &= -3.352 + T + (0.3)(N) \end{aligned}$$

Sliding Block With Friction

and

$$F_{\text{net},y} = W_y + N_y + T_y + F_y$$
$$= (-9.209) + (N) + (0) + (0)$$
$$= -9.209 + N$$

Since we have assumed the acceleration of the hanging mass is down, the acceleration of the sliding mass is to the right at the same rate (this is *implicit data* because they are connected by a rope). Thus,

$$\vec{F}_{\text{net}} = (1.0)(a) = a$$

With the net force aligned with the slope. Therefore,

$$F_{\text{net},x} = -3.352 + T + (0.3)(N) = a$$
$$F_{\text{net},y} = -9.209 + N = 0$$

Collecting the formulas as before, we have

$$T - 19.6 = -2a$$
$$-3.352 + T + (0.3)(N) = a$$
$$-9.209 + N = 0$$

These are similar to the three equations from before, but now the three unknowns are T, N, and a. The last equation yields:

$$-9.209 + N = 0 \Rightarrow N = 9.209$$

Plug this into the first and second equations:

$$T - 19.6 = -2a$$
$$-3.352 + T + (0.3)(9.209) = a$$

Now we have two equations, two unknowns. Let's take the first and solve for T:

$$T = 19.6 - 2a$$

And substitute into the second equation:
$$-3.352 + (19.6 - 2a) + (0.3)(9.209) = a$$
$$\Rightarrow 3a = 19.011$$
$$\Rightarrow a = 6.3369$$

Which is quite fast, about two-thirds g. This seems consistent with the high value of α we derived earlier.

Now, to finish off the problem: "Describe the situation one second later."

Well, we know the acceleration (finally). We also know the initial velocities and the time. We can use Equation (2.6) to get the position and Equation (2.3) to get the final velocities.
$$x = v_i t + \tfrac{1}{2} a t^2 \Rightarrow x = 3.17$$
$$v_f = v_i + at \Rightarrow v_f = 6.34$$

The first mass is being pulled up the slope with an acceleration of 6.34 m/s^2. After one second it has moved 3.17 meters and has a velocity of 6.34 m/s. The second mass is tied to it, so it is falling with the same acceleration with the same velocity and has fallen the same distance.

Circular Motion

The physical problem of an object moving with uniform circular motion carries a certain preeminence. There are a few reasons for this. Historically, going back to the ancient Greeks, the motion of the heavenly objects was connected with the circle. So, the problem has a long pedigree. In addition, it was one of the first physical problems solved in modern times. But more importantly, there are connections with other parts of mechanics: rotation and simple harmonic motion, for example. By laying the foundation of circular motion, we can analyze these more complicated problems in a specific way.

Circular Motion

Newton's First Law states that the natural motion of an object is in a straight line with constant speed.[9] Any deviation from this motion requires an external cause, a force. Given this, when an object moves in a circle, there must be some force maintaining this circular motion. Hooke discovered the mathematical form of this force in the case of uniform circular motion:

$$F = mv^2/r \qquad (2.15)$$

where m is the mass and v is the speed of the object, and r is the radius of the circle. The direction of the force is toward the center of the circle. Because this force always points toward the center of a circle, this force is called *centripetal*. But let's be clear: the centripetal force is *not* a new kind of force, like tension or friction. Any force with the proper magnitude and direction is centripetal whether it is from tension or friction or gravity or whatever.

Stone and Sling

A boy with a 1.0 meter long sling spins a 1.0 kilogram stone horizontally with uniform circular motion. The stone spins one entire revolution in one second. What is the tension in the sling?

Answer: 39 newtons

In this case, the centripetal force is supplied by the tension in the string. Since it is responsible for the uniform circular motion, we have

$$T = mv^2/r = (1.0)(v)^2/(1.0)$$

But what is v? We are told that "the stone spins one entire revolution in one second." This is called the *period* of the

---

[9] This is also known as the principle of inertia.

motion. Any repetitive motion has an associated period: the amount of time required to return to the original state. In this case that occurs after one complete revolution. The period is one second.

But what is v? By definition, it is distance over time. The distance is the circumference of the circle:

$$C = 2\pi r \approx 6.28$$

Since the time involved is one second, the speed is 6.28 m/s. Therefore,

$$T = (1.0)(6.28)^2/(1.0) = 39.478$$

If the boy were to release the sling, the stone would travel with an initial velocity of 6.28 m/s in the direction tangent to the circle.

One of the great explanations of Newton was to identify the gravity of everyday weight with the force that keeps the Moon in orbit. He postulated that the force of gravity decreases with a square of the distance. The Moon orbits at a distance of 384,000 kilometers, which is about 60 times larger than the radius of the earth. So the acceleration due to gravity is 3600 times less than g. That is, $a = 9.80/3600 = 0.00272$ m/s².

Orbital Period of Moon

Calculate the period of the Moon's orbit around the Earth. Assume it orbits with uniform circular motion and an acceleration equal to 0.00272 m/s².

Answer: 2.35×10^6 seconds

The question is "calculate the period". The period is the time it takes to travel one circumference:

$$T = 2\pi r/v$$

Banked Curve

Since we know r, we need to know v. Using Newton's Second Law and the formula for centripetal motion, Equation (2.15), we have:

$$F_{\text{net}} = ma$$
$$\Rightarrow v^2/r = a$$
$$\Rightarrow v = \sqrt{ar} = \sqrt{(0.00272)(3.84 \times 10^8)} = 1022$$

So, the period is

$$T = \frac{2\pi r}{v} = \frac{2\pi(3.84 \times 10^8)}{1022} = 2.3608 \times 10^6$$

This is a little over 27.3 days, which matches observation.[10] This confirmation (and others) led many to accept Newton's theory of gravity as the explanation for the orbits of the heavenly bodies.

Banked Curve

An object need not follow a complete circle in order to apply these concepts.

Banked Curve

Consider an object sliding around a corner without friction. The corner is designed to be at a certain angle θ such that an object moving at 20 m/s slides around a circle of 10 meters. What is the angle?

Answer: 76°

---

[10] This is the "sidereal" month—the amount of time it takes to return to the same point relative to the fixed stars. The "synodic" month (the amount of time from New Moon to New Moon) is 29.5 days.

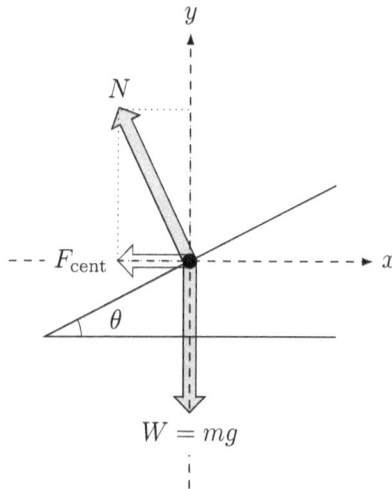

Figure 2.8: Slide banked to produce uniform circular motion

The forces acting are weight and support. But we must be careful with our angles. In this case there are two "natural" choices for the coordinate system: (1) along the slope, or (2) the standard horizontal/vertical directions. The first because of the slope, the second because we know the net force will be horizontal (causing the centripetal motion). Either one will work, but it is better to choose the second.[11] This means that the support force will have components in both x and y directions.

Thus, the weight is

$$\vec{W} = mg \text{ at } 270°$$
$$\Rightarrow W_x = (mg)(\cos 270°) = 0$$
$$\Rightarrow W_y = (mg)(\sin 270°) = -(9.8)(m)$$

---

[11] Try it if you don't believe me. The idea is to align the coordinates with the expected motion of the system.

Banked Curve

and the support force is

$$\vec{N} = N \text{ at } (\theta + 90°)$$
$$\Rightarrow N_x = (N)(\cos(\theta + 90°))$$
$$\Rightarrow N_y = (N)(\sin(\theta + 90°))$$

The components of the net force are

$$F_{\text{net},x} = W_x + N_x = (0) + (N)(\cos(\theta + 90°))$$
$$F_{\text{net},y} = W_y + N_y = -(9.8)(m) + (N)(\sin(\theta + 90°))$$

Where is the centripetal force? *It is the net force.* Remember, "centripetal" describes what a force *does*, not what a force *is*. To maintain the centripctal motion, the net force must have the form:

$$\vec{F}_{\text{net}} = mv^2/r \text{ at } 180°$$
$$\Rightarrow F_{\text{net},x} = -mv^2/r = -(m)(20)^2/(10) = (-40)(m)$$
$$\Rightarrow F_{\text{net},y} = 0$$

Combining these we get:

$$F_{\text{net},x} : (0) + (N)(\cos(\theta + 90°)) = -(40)(m)$$
$$F_{\text{net},y} : -(9.8)(m) + (N)(\sin(\theta + 90°)) = 0$$

Solving the second for N:

$$-(9.8)(m) + (N)(\sin(\theta + 90°)) = 0 \Rightarrow (N) = \frac{(9.8)(m)}{\sin(\theta + 90°)}$$

and substituting out N from the first equation gives

$$\frac{(9.8)(m)}{\tan(\theta + 90°)} = -(40)(m)$$

where we have used the fact that $\tan = \sin/\cos$. We can now solve for the tangent.

$$\tan(\theta + 90°) = -(9.8)/(40) = -0.245$$
$$\Rightarrow \theta + 90 = 166°$$
$$\Rightarrow \theta = 76°$$

Pretty steep slope! But this is because the problem ignores friction. I'll let you work on that problem on your own.

One reason I wanted to add this problem to the set is that, in general, any type of motion can be decomposed into acceleration in the direction of motion and acceleration perpendicular to the motion.[12] The latter component can be considered as caused by an instantaneous centripetal force. This centripetal force has a circle and a radius associated with it. This is how we associate a "radius of curvature" to the curved path.

Ideal Spring

Now for something completely different. Picture a spring, a stiff coil of wire. When it is stretched, the spring responds with a force which pulls back toward equilibrium. The farther you pull, the stronger this force. In addition, if you compress the spring, it responds with an opposing force: again attempting to restore equilibrium. It is an *ideal spring* if the restorative force is exactly proportional to the displacement, i.e.,

$$F = -kx \qquad (2.16)$$

This equation is known as Hooke's Law. The negative sign indicates an opposition to the direction of the displacement (whether it is stretched or compressed). The proportionality constant k is called the *spring constant*. It is different for different springs. Any real spring is ideal for small displacements. For large displacements, Hooke's law will not apply. For even larger displacements, the spring will deform and ultimately break. But for small displacements, it is safe to use the formula.

Hooke's law is our second example of a force which is not con-

---

[12] Doing this is called using the object's "normal" coordinates.

Ideal Spring 53

stant. Technically, the problems involving centripetal force are the first because the direction of that force changes. In this case, the direction does not change, its magnitude does. Although I didn't emphasize it in the previous sections, these forces are not constant so we are not able to use Equation (2.3)–(2.8) to describe the motion.

Also, in the previous section we "cheated" because we started with the motion and figured out the force needed to get that motion. This is the reverse of the procedure we used in the first section and what we need now. Because the force in not constant, we need something new. We have three options:

- Learn calculus.

- Build a numerical solution.

- Use a trick.

All three routes have their merits. Without calculus, normally the only option is the second.[13] However, there is a trick we can use in this case: uniform circular motion.

It is clear that Hooke's Law will create a periodic type of back-and-forth motion. This is similar to circular motion since it is repetitive. But there is an even closer connection: the following argument shows they are actually the same.

The formula for a centripetal force is from Equation (2.15):

$$\vec{F} = -mv^2/r \text{ at } \theta$$

where θ indicates the location of the object on the circle. The negative sign is there because the force points in to the center of the circle. Now, the velocity of the object is given by $v =$

---

[13] What's so bad with the second option? It works every time, but it only works for particular cases. Unless you have a specific problem at hand, having a closed form solution is more useful.

$2\pi r/T$ where T is the period of the motion. If we substitute this in we get:

$$\vec{F} = -m(2\pi r/T)^2/r \text{ at } \theta$$
$$= -(4\pi^2 mr/T^2) \text{ at } \theta$$

What are the components of this force?

$$F_x = -(4\pi^2 mr/T^2)(\cos\theta)$$
$$F_y = -(4\pi^2 mr/T^2)(\sin\theta)$$

And remembering that $x = r\cos\theta$ and $y = r\sin\theta$, we can write:

$$F_x = -(4\pi^2 m/T^2)(x)$$
$$F_y = -(4\pi^2 m/T^2)(y)$$

This shows that both components obey Hooke's Law where

$$k = 4\pi^2 m/T^2 \tag{2.17}$$

So here is the trick. We know the motion that results from uniform circular motion. Now we want to recast that motion in a way that applies to our spring.

The position of the object is defined by the radius of the circle and its angle. Since we are dealing with components, it is more convenient to use x and y coordinates:

$$x = r\cos\theta$$
$$y = r\sin\theta$$

What about time? The motion is *uniform* circular motion, so the angle θ increases at a steady rate. We can represent this as $\theta = \omega t$, where ω is the rate at which the angle changes. This is called the *angular velocity* in an analogy with regular linear velocity. Thus, we have:

$$x = r\cos(\omega t)$$
$$y = r\sin(\omega t)$$

Ideal Spring

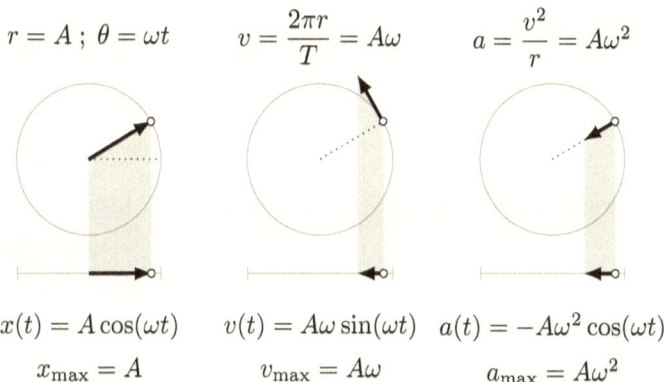

Figure 2.9: Uniform circular motion and simple harmonic motion

The last step is to connect this motion with the force. We want to find a formula that tells us what ω is for a given k. The key is to notice the relationship between T and ω. By definition, ω is the angle turned divided by the time it takes to make the turn. Since T is the amount of time it takes to turn one revolution, we can say that $\omega = 360°/T$. For several reasons it is preferable to use the *radian* as our unit of angle measurement, so we should say:

$$\omega = 2\pi/T \tag{2.18}$$

The radian unit is implied. Finally, if we substitute out the period T from Equation (2.17), we get:

$$\omega = \sqrt{\frac{k}{m}} \tag{2.19}$$

In summary, we have[14]:

$$F = -kx \Rightarrow x = A\cos(\sqrt{k/m}\, t)$$

---

[14] For those of you with some calculus, this may seem like a long, roundabout approach. However, there is more to it than simply avoiding some

Wait! What is this A? This is called the *amplitude* of the motion and depends on how far the spring is initially stretched. Since the cosine function only varies from -1 to 1, the number A represents the maximum displacement experienced by the spring.

That was a long trek. Here is a sample problem to finish this section off.

Ideal Spring – Oscillation Period

A spring hanging vertically is 1.0 meter long. When a 1.0 kilogram mass is attached to the end, it stretches to a length of 1.2 meters. If the mass is slightly displaced, what is period of the oscillation from the spring?

Answer: 0.9 seconds

The question is: "what is the period". This is directly related to the angular velocity by Equation (2.18). To get ω we need to use Equation (2.19). We know the mass, but what is the spring constant? For that we need to use Hooke's Law. What are the forces on the mass? Weight and the spring force. When the spring is stretched to 1.2 meters, it is in equilibrium. At this point, the spring force exactly balances the weight. So we can write:

$$F = -kx \Rightarrow (1.0)(9.8) = -(k)(-0.2) \Rightarrow k = 49.0$$

A note on the signs used here. The spring force is the same size as the weight, but it is positive because it points *up*. The displacement x is measured from equilibrium. It is negative because the displacement is *down*. Notice how this ensures that the spring constant k is a positive number.

---

calculus. This is also introducing us to phasors, without using the term. Phasors will come up again when we talk about optics, AC circuits, and quantum electrodynamics.

Now we use Equation (2.19):

$$\omega = \sqrt{\frac{k}{m}} = \sqrt{\frac{49}{1.0}} = 7.0$$

The period T is related to ω via Equation (2.18):

$$\omega = 2\pi/T \Rightarrow T = 2\pi/(7.0) \Rightarrow T = 0.8976$$

Remember this is the amount of time it takes to cycle back to the *original* state (both position and velocity). This is the amount of time it takes to pass through equilibrium, to the opposite amplitude, pass again through equilibrium, and back to the original displacement. Be careful: some questions may ask for (or give) half or a quarter of the full oscillation period.

Pendulum

The first periodic system to be studied in early modern times was the simple pendulum. Though not a spring, we can use the mathematics of the previous section to our advantage in solving these type of problems.

Pendulum – Oscillation Period

How long must a pendulum be to count one second?

Answer: 0.99 meters

Yikes! Do we know enough to answer this question? A qualified yes. Perhaps the simplest approach to a pendulum is using the notions of torque (which we will discuss in the next chapter), but we can make progress if we make the following assumption: suppose the displacement is small. What is small? Small is such that

$$\sin\theta \approx \theta$$

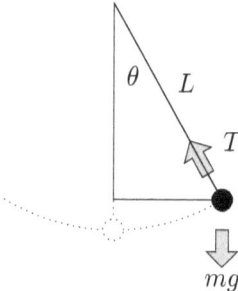

Figure 2.10: Forces on a pendulum

This is usually considered as less than 10 degrees or 0.1 radians. When this applies, we also have

$$\cos\theta \approx 1$$

Suppose the pendulum length is L and has swung out an angle θ. Then the x displacement from equilibrium of the pendulum bob is $L\sin\theta$ and the y displacement is $L - L\cos\theta$. Assuming the angle is small, this reduces to $x = L\theta$ and $y = 0$.

What are the forces on the pendulum? Weight and tension. Since the tension acts as a support force for the pendulum, it must counter-balance the component of the weight in that direction. Thus,

$$T = mg\cos\theta \approx mg$$

and the net force is the remaining component of weight:

$$F_{\text{net}} = mg\sin\theta \approx mg\theta$$

where the approximations hold when the angles are small. Since $x = L\theta$,

$$F_{\text{net}} = mg(x/L)$$

Kepler's Third Law

Which is Hooke's Law with $k = mg/L$. The angular velocity from Equation (2.19) is

$$\omega = \sqrt{\frac{mg/L}{m}} = \sqrt{\frac{g}{L}}$$

So the length controls the period of the pendulum. If we want the clock to tick every time it reaches equilibrium, the period must be *two* seconds. The period is related to the angular velocity via Equation (2.18):

$$T = \frac{2\pi}{\omega} \Rightarrow (2) = \frac{2\pi}{\omega} \Rightarrow \omega = \pi$$

Solving for L yields:

$$\omega = \sqrt{g/L} \Rightarrow (\pi) = \sqrt{(9.8)/L} \Rightarrow L = 0.99$$

Almost exactly one meter! Quite a coincidence.

Here is a trick question: will a large amplitude pendulum keep good time? The answer is yes. The motion will not be exactly sinusoidal like this one because the angle involved is not small—but it will be *periodic*. And that's all you really need for a decent clock.

Kepler's Third Law

We will finish off this chapter with something that is not usually in the beginning textbooks: celestial mechanics. To be fair, we won't get very far either, but I think it's still worth touching. If you want you can skip this section and move on to the next chapter.

Kepler's three laws are as follows:

- The orbits of the planets are ellipses with the Sun at one focus.

- The area swept by the vector from the Sun is constant during the orbit.

- The cube of the orbital period is proportional to the square of the orbital size.

Newton was able to show that these three laws all derive from a single principle: Newton's Law of Gravity. The gravitational force is toward the Sun and has a magnitude equal to:

$$F_{\text{grav}} = \frac{GMm}{r^2} \qquad (2.20)$$

where M is the mass of the sun and G is the gravitational constant equal to 6.673×10^{-11}.

Kepler's Third Law

Assume a planet orbits the Sun in a circular orbit. Using Newton's law of gravity, show that the square of the orbital period is proportional to the cube of the size of the orbit.

Answer: See below

Since the orbit is circular (most of the planets' orbits are only slightly elliptical), the gravitational force must supply the centripetal force: mv^2/r. We can rewrite this in terms of the period T if we remember that $v = 2\pi r/T$. Thus,

$$F_{\text{cent}} = 4\pi^2 mr/T^2$$

Putting this together with Equation (2.20) yields:

$$\frac{4\pi^2 mr}{T^2} = \frac{GMm}{r^2} \Rightarrow MT^2 \propto r^3$$

Which is sometimes called the 1-2-3 law. This shows that Kepler's third law follows from Newton's inverse-square law of gravity. Deriving the other two laws is more difficult.

CHAPTER THREE

MECHANICS II

The previous chapter was focused on the motion of a *particle*. In this chapter, we continue to talk about particles, but we now introduce another layer of description. The ideas of energy and momentum complement the ideas of force and acceleration from Chapter 2. Often the same problems can be solved using either technique. The concepts of energy and momentum are usually easier to use (because there is a conservation law) but not all questions can be answered (usually those concerning time).

Simple Machines and Work Defined

In the previous section about ropes and pulleys we found that we could gain a *mechanical advantage* with a proper configuration of pulleys. We also noted that this advantage was gained at the "expense" of having to pull the rope a larger distance. The same is true of any simple machine.

Mechanical Advantage of Wedge

What is the mechanical advantage of a wedge with an angle of six degrees?

Answer: 10

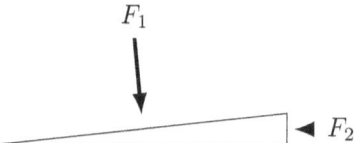

Figure 3.1: Wedge of six degrees

A wedge is basically an inclined plane problem. The force required to push into the inclined plane (call it F_1) is balanced by a much smaller force from the side (call it F_2). The vertical component of F_2 must equal F_1. Thus,

$$F_1 \sin \theta = F_2$$

The input force is from the side (F_2), the multiplied force out the plane (F_1). Thus, the mechanical advantage is:

$$MA = \frac{F_1}{F_2} = \frac{1}{\sin 6°} = 10$$

Thus, a wedge of six degrees will magnify force 10 times. Of course, if we push over a distance, the amount of distance the magnified force pushes is smaller. How much smaller? By the factor $\sin \theta$. Just like with the ropes and pulleys, the mechanical advantage is inversely proportional to distance over which the forces push.

This means that the product of force and distance is the same on the input and the output. This product is called *work*.

Simple Machines and Work Defined

The Lever – Virtual Work

A rigid bar 12 units long has a fulcrum 8 units from the right. On the right is a weight of unknown mass. On the left (4 units from the fulcrum) is a mass of 60 kilograms. Two units to the left of the fulcrum is another mass of 100 kilograms. The whole system is in equilibrium. What is the mass of the unknown weight?

Answer: 55 kilograms

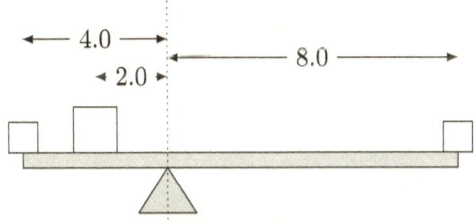

Figure 3.2: A lever problem

Can we use the concept of work in this problem? Yes, in this context it is called the principle of *virtual work*.[1]

Suppose the weight on the right is lifted a small amount, x. Now, we don't claim that the weight is actually lifted this amount, nor even that it *could* be. That is why it is called "virtual". The motion is in the imagination only.

Then the 60 kg mass falls $x/2$ and the 100 kg mass falls $x/4$. The work done by the unknown weight is $-mgx$ (negative because the force is in the opposite direction as the displacement). The work done by the other weights are $(60)(g)(x/2)$ and $(100)(g)(x/4)$, respectively.

---

[1] This is a problem is inspired by a similar problem from *The Feynman Lectures on Physics*, Volume I, Section 4-2.

Since the system is in equilibrium, the total force is zero ... and the total work is zero. Thus,

$$-mgx + (60)(g)(x/2) + (100)(g)(x/4) = 0 \Rightarrow m = 55$$

Notice that the x term cancels in the end: so it doesn't really matter what value you put in, as long as it is small.

Horsepower

Machines are designed to perform work. A good machine does a lot of work. But that is relative. The real question is how much work how fast? In other words, a good machine does a lot of work *quickly*. Thus, we define *power* as the rate at which a machine performs work.

> **Horsepower – Zero to Sixty**
>
> What horsepower is required to accelerate a one ton car from 0 to 60 mph in 4.0 seconds?
>
> **Answer**: 120 horsepower

Since 1 mph = 0.447 m/s, the final speed is 26.82 m/s. The increase in speed occurs over 4 seconds, so the average acceleration is

$$a = \Delta v/\Delta t = (26.82 - 0)/(4) = 6.71$$

Because one ton equals 1016 kg, the force required to create this acceleration is:

$$F = ma = (1016)(6.71) = 6812$$

Since power is the rate at which work is done, we need to calculate work. This requires us to know the total displacement of the car. We need one of our kinematic equations from Chapter 2...

$$x = v_i t + \tfrac{1}{2} a t^2 = (0)(4) + \tfrac{1}{2}(6.71)(4)^2 = 53.68$$

The total work done by the engine is the product of the force and displacement:

$$W = Fx = (6812)(53.68) = 365700$$

This happens in 4 seconds, so the power associated with the work is:

$$P = W/t = 365700/4 = 91400$$

The SI unit for work is the *joule*, and for power it is the *watt*. There are 746 watts in one unit of horsepower, so

$$P = (91400 \text{ J}) \times \frac{1 \text{ hp}}{746 \text{ W}} \Rightarrow P = 122.52 \text{ hp}$$

Projectiles (Again)

In its most basic sense, *energy* is the ability to do work. There are two basic forms of energy: kinetic and potential. Kinetic energy is due to the motion of the parts of a system. When captured by an external agent, this motion can be used to do work. So a motor can be used to lift a weight, move a car, etc.

Potential energy is due to the forces between the parts of a system. An external agent extracts work by changing the configuration of the system. This results in a release of force which is translated into work. A coiled spring and a raised weight are examples.

The formula for translational kinetic energy[2] is

$$KE = \tfrac{1}{2} m v^2 \qquad (3.1)$$

---

[2] Later we will consider *rotational* kinetic energy.

The gravitational potential energy of a particle is

$$PE = mgh \qquad (3.2)$$

The electric potential energy of a charged particle (see Chapter 6) is

$$PE = qV \qquad (3.3)$$

The elastic potential energy of a particle (like a spring) is

$$PE = \tfrac{1}{2}kx^2 \qquad (3.4)$$

The gravitational potential energy from Newton's Law of Gravity is

$$PE = -GMm/r \qquad (3.5)$$

Also, remember that frictional forces do not have potential energies associated with them. They are non-conservative forces because they do not conserve energy. If friction is not present, then the total mechanical energy of the system is conserved. This gives us an alternate way to solve the problems in the previous chapter.

Free-Fall Off Cliff – Using Energy

A car is travelling at 30 m/s horizontally off a cliff 100 meters high. What is its speed when it strikes the ground?

Answer: 53 m/s

It is possible to answer this question using the principles from the previous chapter. We know that the horizontal component of the velocity is unchanged, and the vertical component comes from the acceleration due to gravity. We can use one of the kinematic equations from Chapter 2 to calculate this velocity. And once we know the components of the final velocity, we can calculate the final speed.

However, now we can use energy. There are four steps.

Projectiles (Again)

- Write down the components of the energy at the beginning of the motion. Add them up. This is the initial energy.

- Write down the components of the energy at the end of the motion. Add them up. This is the final energy.

- The difference between the final and initial energy is due to either an external force or friction. If neither is present, the difference is zero. Write this down as an equation: $\Delta E = 0$, for example.

- Plug in the initial and final energy into the equation and use it to answer the question.

We start with the initial state.

$$KE_i = \tfrac{1}{2}mv_i^2 = \tfrac{1}{2}(m)(30)^2 = (450)(m)$$
$$PE_i = mgh_i = (m)(9.8)(100) = (980)(m)$$

The total initial energy is therefore

$$E_i = (1430)(m)$$

Now for the final state.

$$KE_f = \tfrac{1}{2}mv_f^2 = \tfrac{1}{2}(m)(v_f)^2$$
$$PE_f = mgh_f = (m)(9.8)(0) = 0$$

The total final energy is

$$E_f = \tfrac{1}{2}(m)(v_f)^2$$

In this problem, the change in energy is zero. Friction and air drag are ignored (as usual) and there are no external forces (gravity is taken into account by potential energy).[3] Thus,

$$\Delta E = E_f - E_i = 0$$
$$\Rightarrow \tfrac{1}{2}(m)(v_f)^2 - (1430)(m) = 0$$
$$\Rightarrow v_f = 53.479$$

Here is another projectile problem. This one shows up in many different forms. An example is the bottle rocket problem from the previous chapter.

Free-Fall – How Fast From Duration – Using Energy

A bottle rocket is launched straight up. It reaches a maximum height of 20 meters. How fast was it launched?

Answer: 20 m/s

Initially, the potential energy is zero, so the total initial energy is just $E_i = \tfrac{1}{2}mv_i^2$. At the maximum height the velocity is zero (remember *implicit data*), so there is no kinetic energy, therefore the total final energy is

$$E_f = mgh = (m)(9.8)(20) = (196)(m)$$

Since the change in energy is zero, we have

$$(196)(m) - \tfrac{1}{2}mv_i^2 = 0 \Rightarrow v_i = 19.799$$

This method is quite a bit easier than the solution from the previous chapter. They both get to the same answer, as they should. Don't worry if you don't choose the most "elegant" way to solve a problem. As long as your logic is right, you will get the right answer.

Here is a similar problem:

---

[3] Technically, the system is both the car and the Earth. This is why the gravitational force is *internal* rather than external. Notice that the height is a measure the distance between the two parts of the system.

A Roller Coaster

> **Skateboard Quarter-Pipe – Using Energy**
>
> A skateboarder is travelling at 10 m/s. He rides a quarter-pipe which launches him in the air. How high does he get?
>
> **Answer**: 2.6 meters

Initially,

$$KE_i = \tfrac{1}{2}mv_i^2 = (50)(m)$$
$$PE_i = mgh = 0$$

In the final moment,

$$KE_f = 0$$
$$PE_f = mgh$$

There is no change to mechanical energy, so

$$\tfrac{1}{2}(50)(m) = (m)(9.8)(h) \Rightarrow h = 2.5510$$

These energy problems can be quite easy once you get the hang of them.

A Roller Coaster

The following problem emphasizes that the kinetic energy of an object only depends on its speed, and therefore does not take into account direction.

> **The Roller Coaster and The Coin**
>
> A child on a roller coaster accidentally drops a coin just as it reaches the top of an incline 50 meters high. The roller coaster is barely moving (let $v_i = 0$). Assume no friction and no air drag. What is the speed of the roller coaster when it reaches the ground before the brakes are applied? How does this compare to the speed of the coin when it strikes the ground?
>
> **Answer**: They are the same.

Since there is no kinetic energy at the beginning, the initial mechanical energy of the coin is

$$mgh = (490)(m)$$

The final energy of the coin is $\frac{1}{2}mv_f^2$ because there is no potential energy at the end. Since there is no friction, the change in energy is zero. Therefore

$$\Delta E = 0 \Rightarrow (490)(m) - \tfrac{1}{2}mv_f^2 = 0 \Rightarrow v_f = 31.305$$

Similarly, the initial energy of the roller coaster is

$$Mgh = (490)(M)$$

And the final energy is $\frac{1}{2}Mv_f^2$. Since there is no friction, the energy should not change.

But wait! What about all the support forces on the roller coaster? It loops and twists and turns. These do not change the energy either. Why? They all act perpendicular to the motion. They change the *direction* of the velocity, but not the speed. Any constraint force behaves in this way (the magnetic force does this too). They will not change the mechanical energy of the system. Of course, in the real world, such forces of constraint involve a lot of friction, but we choose to ignore that fact.

So, the logic is the same for the roller coaster and the coin:

$$\Delta E = 0$$
$$\Rightarrow (490)(M) - \tfrac{1}{2}Mv_f^2 = 0$$
$$\Rightarrow v_f = 31.305$$

The direction of the motion is different, but the kinetic energy is the same.

Energy Lost To Drag

Let's talk friction.

> ### Calculating Friction – Bullet in Wood
>
> A 0.010 kilogram bullet is fired at 200 m/s and comes to rest lodged 0.10 meters into a block of wood (fixed to the ground). What is the average force of friction in the block bringing the bullet to rest?
>
> **Answer**: 2000 newtons

The average force can be calculated from the work done, via $W = Fx$. Since we know the distance over which this force acts, if we knew the work done we could answer the problem. The work done by friction is what changes the mechanical energy of the bullet. The initial energy is

$$\tfrac{1}{2}mv^2 = \tfrac{1}{2}(0.010)(200)^2 = 200$$

Since the energy of the bullet is zero at the end of the problem, this is the work done by the average friction. Thus,

$$W = Fx \Rightarrow (200) = (F)(0.10) \Rightarrow F = 2000$$

Turning Points

Turning points are a generalization of the projectile problems dealing with the maximum height. Suppose one plots the potential energy of, say, a spring using Equation (3.4). Draw a horizontal line representing the total energy of the system. Regardless of where the system is, the total energy is at this constant level. The kinetic energy of the system is the difference between these two lines.

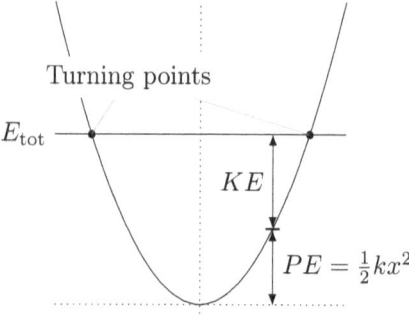

Figure 3.3: The energy diagram efficiently summarizes the dynamics of a system.

Since the kinetic energy must be positive, the system cannot occupy the places where the potential exceeds the total energy. The points where the lines intersect are called *turning points*. These represent the moment when the kinetic energy drops to zero and the system turns around. The following is a problem that uses the turning point idea.

Turning Points

Gravitational Turning Point

An object with a mass of 1000 kilograms is in the gravitational grip of the Sun (mass = 1.99×10^{30} kilograms). At a distance of 5.00×10^{11} meters, the object has a velocity of 5000 m/s directly away from the Sun. What is the maximum distance it will travel before it is pulled back?

Answer: 5.24×10^{11} meters

As I mentioned, the solution is very similar to the projectile problem. What is the initial kinetic energy of the object?

$$KE_i = \tfrac{1}{2}mv_i^2 = \tfrac{1}{2}(1000)(5000)^2 = 1.25 \times 10^{10}$$

And the initial potential energy?

$$PE_i = -GMm/r_i$$
$$= -(6.673 \times 10^{-11})(1.99 \times 10^{30})(1000)/(5.00 \times 10^{11})$$
$$= -2.67 \times 10^{11}$$

So the total energy is

$$E = 1.25 \times 10^{10} - 2.67 \times 10^{11} = -2.545 \times 10^{11}$$

The final kinetic energy is zero. The final potential energy is:

$$PE_f = -GMm/r_f$$
$$= -(6.673 \times 10^{-11})(1.99 \times 10^{30})(1000)/(r_f)$$

Since the total energy is the same in the initial and final states, we have

$$-2.545 \times 10^{11} = -(6.673 \times 10^{-11})(1.99 \times 10^{30})(1000)/(r_f)$$
$$\Rightarrow r_f = 5.24 \times 10^{11}$$

> **Ideal Spring – Amplitude**
>
> A spring with a spring constant of $k = 5.0$ N/m is attached to a 2.0 kilogram mass. The system is in equilibrium. Then, in a moment, the mass is given a velocity of 0.40 m/s. What is the maximum amplitude of the resulting oscillation?
>
> **Answer**: 0.25 meters

Again we ask about the initial energy. Since there is no potential energy, it is all kinetic:

$$\tfrac{1}{2}mv_i^2 = \tfrac{1}{2}(2.0)(0.40)^2 = 0.16$$

At the end of the motion, the kinetic energy is zero ("maximum amplitude"), so all of this is potential energy. In other words,

$$PE_f = \tfrac{1}{2}kx_f^2 = \tfrac{1}{2}(5.0)(x_f)^2 = 0.16 \Rightarrow x_f = 0.25298$$

Collisions And Momentum

Now for something completely different: momentum.

Suppose we have two particles. They may interact somehow between themselves, but assume there are no external forces present. Because of the interaction, they may move in an intricate pattern, but not their center of mass. Only external forces can accelerate the center of mass. But what is the center of mass?

The center of mass is calculated using

$$x_{\text{cm}} = \frac{m_1 x_1 + m_2 x_2}{m_1 + m_2} \tag{3.6}$$

This is the point in space where that action of the object is "concentrated".

Collisions And Momentum

> **Center of Mass**
>
> Calculate the center of mass between the Earth and the Moon. Their masses are 5.98×10^{24} and 7.36×10^{22} kilograms respectively. The distance between them is 3.84×10^8 meters.
>
> **Answer**: 4.67×10^6 meters

If we call the location of the Earth $x_1 = 0$, then the location of the Moon is $x_2 = 3.84 \times 10^8$. Thus,

$$\begin{aligned} x_{\text{cm}} &= \frac{m_1 x_1 + m_2 x_2}{m_1 + m_2} \\ &= \frac{(5.98 \times 10^{24})(0) + (7.36 \times 10^{22})(3.84 \times 10^8)}{(5.98 \times 10^{24}) + (7.36 \times 10^{22})} \\ &= 4.67 \times 10^6 \end{aligned}$$

Momentum is the product of mass and velocity. The total momentum of a system is conserved when no external forces act upon it. Understanding the momentum of a system is a necessary compliment to the energy of a system. For example: suppose two objects of equal mass are moving with equal and opposite velocities. They collide and combine. By symmetry, the final velocity of the system must be zero. The final kinetic energy is zero, even though there is some initial kinetic energy. Energy is not conserved.

On the other hand, momentum *is* conserved. It was zero before (because momentum takes into account direction) and it is still zero after. Momentum is always conserved in any collision.

Consider the following problem.

> **Completely Inelastic Collision**
>
> Particle #1 has a mass of 6.0 kilograms and is moving to the right with speed 20 m/s. Particle #2 has a mass of 4.0 kilograms and is moving to the left with speed 5.0 m/s. They collide and combine. What is the final velocity of the aggregate?
>
> **Answer**: 10 m/s

Though different than energy, we approach momentum problems in the same way. What is the initial momentum?

$$m_1 v_{i1} = (6.0)(20) = 120$$
$$m_1 v_{i2} = (4.0)(-5.0) = -20$$

In total, the initial momentum is $(120 - 20) = 100$ kg-m/s. What is the final momentum?

$$m_1 v_{f1} = (6.0)(v_f)$$
$$m_1 v_{f2} = (4.0)(v_f)$$

Because they collide and combine, the final velocities are the same. In total, the final momentum is $(6.0)(v_f) + (4.0)(v_f) = (10)(v_f)$. Since there is no external force, these two totals are equal to one another.

$$100 = (10)(v_f) \Rightarrow v_f = 10$$

The positive sign shows that the motion is to the right.

These types of collisions are a special type of *inelastic collision*. They involve two objects that stick together in the end. The next problem is an example of a different kind of collision. It is basically an inelastic collision in reverse. We will use the same principles to solve it.

Collisions And Momentum

Explosive Disintegration

The Apollo spacecraft is composed of two parts: the command/service module (mass = 30,300 kg) and the lunar module (mass = 14,700 kg). The ship is at rest relative to its center of mass. Then the modules separate. The speed of the modules relative to one another is 10 m/s. What are the speeds of the two modules relative to the center of mass?

Answer: 3.27 m/s and −6.73 m/s.

The question is "What are the speeds". It never hurts to give them names. Let's call the command/service module #1 and the lunar module #2. We are looking for both v_{f1} and v_{f2}. The initial momentum is zero since they are both at rest. Since momentum is conserved, the final momentum is also zero. Thus,

$$(30300)(v_{f1}) + (14700)(v_{f2}) = 0$$

We have one equation, two unknowns. We need another equation. It is in the statement "The speed of the modules relative to one another is 10 m/s." In symbols:

$$v_{f1} - v_{f2} = 10$$

Now we have two equations, two unknowns. From the second equation we have $v_{f1} = v_{f2} + 10$ which we can substitute into the first equation and solve:

$$(30300)(v_{f2} + 10) + (14700)(v_{f2}) = 0$$
$$\Rightarrow (45000)(v_{f2}) = -303000$$
$$\Rightarrow v_{f2} = -6.7333$$

Therefore,

$$v_{f1} = v_{f2} + 10 \Rightarrow v_{f1} = 3.2667$$

This makes sense. The velocities are in opposite directions. The larger mass has the smaller velocity.

So far, we've kept it simple. The two objects have been in tandem either at the beginning or the end of the collision. In both cases kinetic energy was not conserved. When the collision is such that kinetic energy *is* conserved, this is called an *elastic collision*.

> **Elastic Collision**
>
> A ball with a mass of 10 kilograms is initially at rest. It is struck by a smaller ball with mass 1.0 kilogram moving at 5.0 m/s to the right. The collision is elastic. What are the final speeds?
>
> **Answer**: 0.91 m/s and -4.1 m/s

Since the collision is elastic, the problem involves *both* the conservation of momentum and the conservation of energy. We start with the initial state. What is the initial momentum?

$$m_{i1}v_{i1} + m_{i2}v_{i2} = (10)(0) + (1.0)(5.0) = 5.0$$

What is the initial kinetic energy?

$$\tfrac{1}{2}m_{i1}v_{i1}^2 + \tfrac{1}{2}m_{i2}v_{i2}^2 = \tfrac{1}{2}(10)(0)^2 + \tfrac{1}{2}(1.0)(5.0)^2 = 12.5$$

Okay. Now the final state. What is the final momentum?

$$m_{f1}v_{f1} + m_{f2}v_{f2} = (10)(v_{f1}) + (1.0)(v_{f2})$$

What is the final kinetic energy?

$$\tfrac{1}{2}m_{i1}v_{i1}^2 + \tfrac{1}{2}m_{i2}v_{i2}^2 = (5.0)(v_{f1})^2 + (0.5)(v_{f2})^2$$

The initial momentum equals the final momentum and the initial kinetic energy equals the final kinetic energy. This gives us two equations for two unknowns:

$$5.0 = (10)(v_{f1}) + (1.0)(v_{f2})$$
$$12.5 = (5.0)(v_{f1})^2 + (0.5)(v_{f2})^2$$

The first is the easier one this time. We have
$$v_{f2} = 5.0 - (10)(v_{f1})$$
Substitute this into the second and we get
$$12.5 = (5.0)(v_{f1})^2 + (0.5)(5.0 - (10)(v_{f1}))^2$$
Now the gymnastics begin...
$$12.5 = (5.0)(v_{f1})^2 + (0.5)(5.0 - (10)(v_{f1}))^2$$
$$\Rightarrow 12.5 = (5.0)(v_{f1})^2 + (12.5 - (50)(v_{f1}) + (50)(v_{f1})^2)$$
$$\Rightarrow 0 = -(50)(v_{f1}) + (55)(v_{f1})^2$$
$$\Rightarrow 50 = (55)(v_{f1})$$
$$\Rightarrow v_{f1} = 0.9091$$

The 10 kilogram mass initially at rest is now moving at 0.91 m/s to the right. And the other mass?
$$v_{f2} = 5.0 - (10)(v_{f1})$$
$$\Rightarrow v_{f2} = 5.0 - (10)(0.91)$$
$$\Rightarrow v_{f2} = -4.0909$$

The 1.0 kilogram mass *rebounds* and is now moving at 4.1 m/s to the left.

Ballistic Pendulum

Ballistic Pendulum

A 0.010 kilogram bullet is fired into a 1.0 kilogram block of wood. The wood is suspended from the ceiling by a thin rope and swings to a maximum height of 0.25 meters. What is the initial velocity of the bullet?

Answer: 220 m/s.

It's tempting to think through the problem the following way. The initial kinetic energy of the bullet is

$$\tfrac{1}{2}mv^2 = \tfrac{1}{2}(0.010)(v)^2$$

and the initial kinetic energy of the block is zero. Since the final potential energy of the both is

$$mgh = (0.010 + 1.0)(9.8)(0.25) = 2.475$$

the velocity of the bullet was

$$\tfrac{1}{2}(0.010)(v)^2 = 2.475 \Rightarrow v = 22$$

But this is *wrong*.

The issue is that the collision of the bullet with the block of wood is inelastic. Kinetic energy is *not* conserved start to finish. This is actually a two phase problem: (1) the inelastic collision; and (2) the pendulum-like swing.

In the first phase we can use the conservation of momentum and in the second phase we can use the conservation of energy. Connecting the two we have the velocity of the bullet/block combination. It is used in the final momentum of phase one and the initial kinetic energy of phase two.

If we start with the first phase, we are using the conservation of momentum:

$$(m_1 v_{i1} + m_2 v_{i2}) = (m_1 + m_2)(v_f)$$
$$\Rightarrow (0.010)(v_{i1}) + (1.0)(0) = (0.010 + 1.0)(v_f)$$

Hmm. One equation, two unknowns. What can we do? Move on to the second phase. We are using the conservation of energy. At the beginning of the phase there is no potential energy and at the end there is no kinetic energy. Since there is no energy lost *during the swing*, we can say:

$$KE_i = PE_f$$
$$\Rightarrow \tfrac{1}{2}(m_1 + m_2)(v_f)^2 = (m_1 + m_2)(g)(h_f)$$
$$\Rightarrow \tfrac{1}{2}(1.01)(v_f)^2 = (1.01)(9.8)(0.25)$$
$$\Rightarrow v_f = 2.213$$

Good. No "one equation, two unknowns" thing. Plug this into the first equation.

$$(0.010)(v_{i1}) = (1.01)(2.213) \Rightarrow v_{i1} = 223.51$$

That is a much more reasonable speed for a bullet.

Rocket Propulsion

I include this problem not because it is standard, but because it is interesting and usually ignored. Jump to the next section is you are strapped for time. Since one needs some calculus to do the real problem, I've changed it.

Rocket Propulsion With Two Apples

A kid (mass = 10.0 kilograms) on roller skates holds two apples (0.100 kilograms each). He can either throw both apples together or throw them in succession to gain maximum recoil velocity. In either case, the speed at which he throws the apples is the same: 10.0 m/s. Which should he do and what is the speed differential?

Answer: Throw them together; the recoil speed is −0.200 m/s

Since the kid throws the apples at same speed every time and they all have the same mass, the momentum of each is the same. Since there are no external forces momentum is conserved. (This is why he is "on roller skates"—you are supposed to ignore friction). The kid's body recoils from the momentum in the apple: he gains momentum backward.

If he throws one apple and holds the other, some of the recoil momentum is shared by the apple he holds and his speed differential is less than one-half of what he would gain by throwing them both. So that's bad. Also, when he throws the last apple, since that apple is moving slightly, his speed differential is less to one-half of what he would gain by throwing both. Double bad. He should throw them both to gain maximum speed.

Here are the calculations. Suppose he throws both. The momentum in the thrown apples is

$$2mv = (2)(0.100)(10.0) = 2.00$$

Momentum is conserved, so his recoil momentum is -2.00 kg-m/s. Since his mass is 10.0 kilograms, his recoil speed is -0.200 m/s.

Now suppose he throws them one after the other. The momentum of the first apple is

$$mv = (0.100)(10.0) = 1.00$$

Momentum is conserved, so his recoil momentum is -2.00 kg-m/s. Since his mass is 10.1 kilograms, his recoil speed is -0.099 m/s. He throws the second apple. Since he throws it at 10.0 m/s while moving at -0.099 m/s, the ground speed of the second apple is 9.90 m/s. Therefore, the "ground momentum" is 0.990 kg-m/s. The kid acquires this in recoil: his speed differential is -0.099 m/s.[4] The total speed differential is -0.198 m/s.

---

[4] It is a coincidence that these two speed differentials are the same. If you do the same calculation for three apples, this pattern does not hold.

This total speed differential is called *delta-v* (i.e., Δv) in rocket propulsion. The formula for the total delta-v of a rocket of mass M with propellant of mass m ejected at a speed u is

$$\Delta v = -u \ln(1 + m/M)$$

This assumes the propellant is burned continuously. If the propellant were ejected all at once at that speed, then $\Delta v = -u(m/M)$.

Rain Pressure

Consider a force that acts over a very short time period (e.g., a bat hitting a baseball). The average value of the force multiplied by the duration in which it acts is called the *impulse* of the force. Impulse changes momentum. This is really just Newton's Second Law again.

$$F \Delta t = \Delta(mv) \Leftrightarrow F = \Delta(mv)/\Delta t = ma$$

The following problem is an application of the impulse idea, but it also is preparation for a key step in deriving the internal energy of a an ideal gas in Chapter 5.

Rain Pressure

What is the average force felt from a machine gun firing 0.010 kilogram bullets at 50 bullets per second? The bullet speed is 200 m/s.

Answer: 100 newtons

Sometimes a problem includes an unusual rate (not velocity or force or something) like this one: "50 bullets per second". I have found it is much easier to think about the problem by rewording it in the following way:

"What is the average force felt from a machine gun firing 0.010 kilogram bullets at 50 bullets in one second? The bullet speed is 200 m/s."

This makes explicit the time period involved. Of course the problem is the same if one were to say "100 bullets in two seconds". So this time period is arbitrary, but I find this makes using the equations easier.

The reason we are thinking about using impulse here is that it is a combination of force and time. We are looking for the "average force". This phrase also comes up in problems dealing with work, so look for applying those equations when you see these keywords also. But, in this case, we use impulse.

$$F \Delta t = \Delta(mv)$$
$$\Rightarrow (F)(1) = (0.010)(200) - 0$$
$$\Rightarrow F = 2.00$$

This is the average force from each bullet. To get the total force we need to multiply by the number of bullets:

$$F_{\text{tot}} = (n)(F_{\text{avg}}) = (50)(2.00) = 100$$

I have assumed the bullets become lodged and come to rest (final momentum equal to zero). What if the bullets recoil? Suppose they recoil with a velocity of -10 m/s. Then the answer becomes:

$$F \Delta t = \Delta(mv)$$
$$\Rightarrow (F)(1) = 50((0.010)(200) - (0.010)(-10))$$
$$\Rightarrow F = 105$$

When recoil is involved the average force is larger.

CHAPTER FOUR

MECHANICS III

Now we move on to things that are not particles. This includes

- Rigid bodies
- Deformable solids
- Fluids
- Wave motion

Rigid bodies have extension, so they can rotate. When we talk about solids and fluids, however, the fundamental equations are much more complicated. In an introductory class, one is usually only exposed to a few basic principles. We will cover the relevant problems in this chapter.

In addition, wave motion is characteristic of both solid and fluids. The applications to sound, light, and quantum mechanics make it particularly important to at least introduce the basic concepts of wave motion.

Tension And Torque

When discussing the lever last chapter, we used the principle of virtual work. There is an alternative way to solve these problems using *torque*. Torque is the twisting action of a force around a specific point in space. Not only the distance between the specific point and where the force is applied is important, but also the angle of the force. These facts yield a new concept: the *lever arm*. To find the lever arm of a force, one imagines drawing a line in the direction of the force through its point of application. The shortest distance between the specified point and this line is the level arm. This distance is measured perpendicular to the force line. The torque of a particular force is the magnitude of the force multiplied by its lever arm (ℓ).

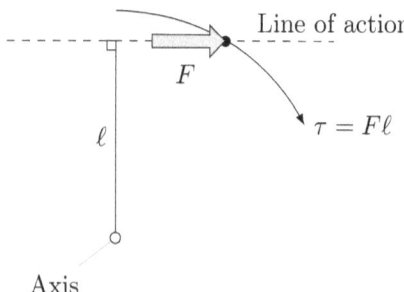

Figure 4.1: Torque is the rotational analog of force and is defined by a line of action and a lever arm.

In this chapter, we are no longer dealing with particles. We need to pay attention to *where* the forces are applied on these objects. Typically this is clear, but where should we consider the weight of an extended object to be applied?

The answer is: in its center. But it's not always that simple. In order to explain the issue, I would like to distinguish three different "centers". In general, they are all in the same spot.

Tension And Torque

The three centers are:

- The geometric center
- The center of mass
- The center of gravity

The geometric center is obvious for simple shapes: a circle, a square, etc. For a more complicated shape without any symmetry there are rules to find it (pure geometry or calculus), that I won't get into it now. The center of mass is the point at which an object can be considered concentrated. We discussed this in the previous chapter. The center of gravity is similar: it is the point where an object's weight can be considered concentrated. It is the "average" of all the parts of the object with the average weighted by their weight.

For example: imagine a lever with ten kilograms on one end and five kilograms that is six meters away. The geometric center is in the middle, three meter away. But the center of gravity is closer to the ten kilogram mass: two meters away. This is the same calculation as the center of mass in the previous chapter, but with the mass replaced with weight:

$$x_{\text{cm}} = \frac{W_1 x_1 + W_2 x_2}{W_1 + W_2} \qquad (4.1)$$

One can take any force and calculate its respective center. But since gravity is a constant force (on the surface of the earth), every g from $W = mg$ cancel, so there is no difference between the center of gravity and the center of mass.

The point is that the center of gravity is usually in the geometric center of the object. The two requirements for this to be true are: (1) the object is of uniform density; and (2) the force of gravity is constant. It is usually safe to make these two assumptions unless the problem states otherwise.

88　Chapter 4.　Mechanics III

> **Tension Required To Support Hanging Sign**
>
> A sign 2.0 meters long and 1.5 meters high hangs perpendicular to the wall of a building to the right. The top corner closest is pinned to the wall and the far top corner is attached to a wire which is also attached to the building 1.0 meter up. If the mass of the sign is 10 kilograms, what is the tension in the wire?
>
> **Answer**: 110 newtons

This can be solved using the principle of virtual work (imagine the length of the wire changing), but we choose here to use torque.

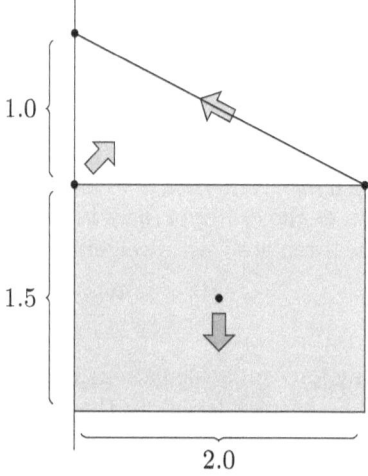

Figure 4.2: Forces on a hanging sign

Okay. We have a rectangular sign. The tension is pointing up and to the left from the top-right corner. The support is

Tension And Torque

pointing up[1] and to the right from the top-left corner.[2] Gravity is pointing straight down from the center of the sign. Now what?

We need to calculate the torques. They all counter-balance because the sign is in equilibrium. From where do we calculate them? The answer is: anywhere. Since there is no rotation, there is no obvious point to use. The torques around any point we choose must cancel.

We choose to calculate the torque around the top-left corner. Why? This is the force we know the *least* about. We know neither its magnitude nor its direction. But if we choose this point, its level arm is zero, effectively removing it from the torque calculation.

Now the hard work. What is the torque of the tension? We need to find the lever arm. Prepare for some trig. Draw a triangle with the base along the wire (this is the line of force) and drop a perpendicular to the top-left corner. This is the lever arm. The hypotenuse of the triangle is the top of the sign. (See Figure 4.3)

We have a triangle, where is the angle? Focus on the angle between the wire and the sign. Since the wire is attached 1.0 meters up and 2.0 meters away, this angle is given by

$$\tan\theta = 1.0/2.0 \Rightarrow \theta = 26.6°$$

On the other hand, the lever arm is in a right triangle with a

---

[1] How do I know it is up? I don't. If I guess wrong, the final sign on the vertical component will tell me.

[2] You might wonder why this support force is not "normal" to the support surface like those in Chapter 2. The reality is that every support is actually a manifestation of the elastic forces of the material providing the support. In this case the elastic force has both normal and shear components—but we have not introduced this concept yet! For now we will leave this support force TBD.

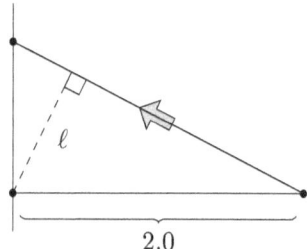

Figure 4.3: The lever arm of the tension on the hanging sign

2.0 meter hypotenuse. Therefore:

$$\sin 26.6° = \ell/2.0 \Rightarrow \ell = 0.894$$

The lever arm of the tension is 0.894 meters.

Now calculating the torque is easy: $\tau = F\ell = (0.894)(T)$. But we don't know T. Moving on.

What is the torque of the weight? Again, we need to find its lever arm. In this case it is easy. The line of force is directly vertical, so the lever arm is 1.0 meter. The resulting torque is

$$\tau = (mg)\ell = (10)(9.8)(1.0) = 98$$

Since the total torque is zero, we can say

$$\tau_T + \tau_W = 0 \Rightarrow (0.894)(T) - 98 = 0 \Rightarrow T = 109.62$$

Did I forget to mention the sign of the torque? If the force is aligned with a counter-clockwise motion along the line of force the torque is positive, otherwise negative.

Let's summarize the procedure:

- Identify all the forces and their points of application.

- Choose a point of rotation. If the situation is in equilibrium, you are free to choose any point. Choose one on top

Double Support

of the force you know the least about (which eliminates its torque from the problem).

- Calculate the lever arm for each force. Do this by drawing the line of action. The line perpendicular to this line through the point of rotation is the lever arm. This usually involves a mess of trig.

- Calculate the torque for each force. This is just the product of the magnitude of the force and its lever arm. The sign of the torque is positive if the force points along the line of action counter-clockwise.

- Add all the torques. If the system is in equilibrium they must add to zero. If this is not enough information, use the fact taht the linear components of all the forces must add to zero also.

- Finally, answer the question.

In our case, we didn't need to use the components of the forces. But, for example, if we were asked the direction of the support force we would need all three equations. One torque equation and two force component equations.

Double Support

These type of problems are similar to the one above.

Double Support

A table has a 100 kilogram mass on it. The 50 kilogram table is 2.0 meters long and the mass is 0.10 meters from the left edge. What are the support forces in the four legs?

Answer: 1200 and 290 newtons

There are six forces in this problem.

- Two support forces from the legs on the left side, straight up.

- Two support forces from the legs on the right side, straight up.

- One force from the weight of the table in the middle of the table: 1.0 m from each edge, straight down. $W = mg = (50)(9.8) = 490$.

- One from the weight of the extra mass—at 0.10 m from the left edge, straight down. $W = mg = (100)(9.8) = 980$.

As for the point of rotation, we should choose one of these points of application. Let's choose the left edge.

Fortunately, finding the lever arm for each of these forces is simple.

- Two support forces on the left: $\ell = 0.0$

- Two support forces on the right: $\ell = 2.0$

- The weight of the table: $\ell = 1.0$

- The weight of the extra mass: $\ell = 0.10$

The resulting torques are:

Flywheel

- Two support forces on the left: $\tau = 0$
- Two support forces on the right: $\tau = (N_2)(2.0)$
- The weight of the table: $\tau = -(50)(9.8)(1.0) = -490$
- The weight of the extra mass: $\tau = -(100)(9.8)(0.10) = -98$

The total torque is:

$$(N_2)(2.0) - 490 - 98 = 0 \Rightarrow N_2 = 294$$

We could find the other support force by starting over with another rotation point. We can also use the components of the forces. Since the forces are all vertical, we have:

$$(N_1) - (50)(9.8) - (100)(9.8) + (294) = 0 \Rightarrow N_1 = 1176$$

Round to two significant digits and you are done.

Flywheel

We now move on to problems that are not in equilibrium. A net torque produces rotation. The formula is

$$\tau = I\alpha \qquad (4.2)$$

This is Newton's Second Law applied to rigid body rotation. The letter I represents the *moment of inertia* for the object. The moment of inertia for a particle is $I = mr^2$. The moment of inertia for a particular geometric object (sphere, rod, etc.) is something you need to look up in a table. The letter α represents the angular acceleration of the object. Notice that the angular velocity has a sign, just like torque.

Flywheel and Pulley

A horizontal disk with mass 2.0 kilogram has a radius of 0.10 meters. Around the circumference is a string that runs over a pulley supporting a 0.25 kilogram mass. From rest, how long does it take the mass to fall 1.0 meter?

Answer: 1.0 second

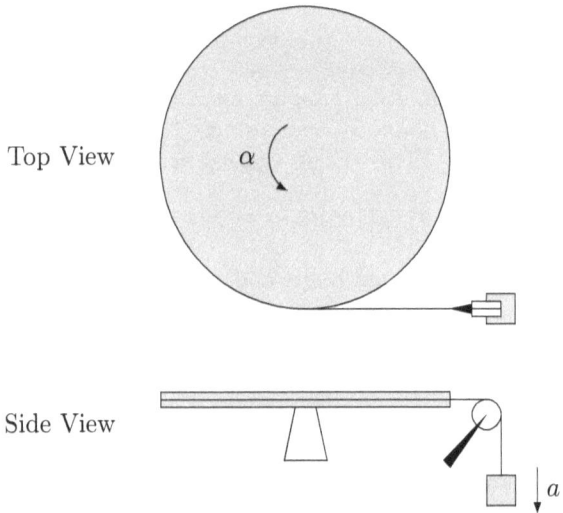

Figure 4.4: Flywheel and pulley

This is a set up you might encounter in a lab. Let's start with the question: "how long". This part of the problem is from Chapter 2. We need to know the acceleration of the mass. From the kinematic equations in Chapter 2 we know that

$$x = \tfrac{1}{2}at^2$$

This is also the acceleration of the rim of the disk since this is where the string is. The acceleration of the rim of the disk is

Flywheel

related to the angular acceleration via

$$a = r\alpha$$

And the angular acceleration we can calculate from Equation (4.2) if we knew the torque. The torque comes from the tension in the string supporting the hanging mass:

$$\tau = (T)(\ell)$$

The lever-arm is the radius of the disk since the string is pulling from the rim. However, we cannot assume the tension equals the weight of the hanging mass since the system is not in equilibrium.[3] In fact, we have

$$T - mg = -ma \Rightarrow T = mg - ma$$

This string of equations involves all the information we have. Let's plug in the data. The tension:

$$T = (0.25)(9.8) - (0.25)(a) = 2.45 - (0.25)(a)$$

We don't know a yet, so we leave it in until we do. Now the torque:

$$\tau = (2.45 - (0.25)(a))(0.10) = 0.245 - (0.025)(a)$$

The moment of inertia for the disk is:

$$I = \tfrac{1}{2}mr^2 = \tfrac{1}{2}(2.0)(0.1)^2 = 0.01$$

The angular acceleration is:

$$\tau = I\alpha$$
$$\Rightarrow (0.245 - (0.025)(a)) = (0.01)(\alpha)$$
$$\Rightarrow \alpha = 24.5 - (2.5)(a)$$

---

[3] Be careful! This is something that is frequently missed by students.

The linear acceleration at the rim of the disk is:

$$a = r\alpha = (0.10)(24.5 - (2.5)(a))$$

We can solve for a:

$$a = 2.45 - (0.25)(a) \Rightarrow a = 1.96$$

Finally, the duration of the fall is given by:

$$x = \tfrac{1}{2}at^2$$
$$\Rightarrow (1.0) = \tfrac{1}{2}(1.96)(t)^2$$
$$\Rightarrow t = 0.9899$$

If the mass were not attached to the flywheel, the duration would be 0.45 seconds.

We will now repeat this question, but instead we solve it using energy. The formula for the kinetic energy due to rotation is:

$$KE = \tfrac{1}{2}I\omega^2 \qquad (4.3)$$

Flywheel and Pulley – Use Energy

A horizontal disk with mass 2.0 kilograms has a radius of 0.10 meters. Around the circumference is a string that runs over a pulley supporting a 0.25 kilogram mass. From rest, how long does it take the mass to fall 1.0 meter? Use the conservation of energy to solve this problem.

Answer: 1.0 second

We need to somehow use the formula for rotational kinetic energy, Equation (4.3). Initially, ω is zero and the velocity of the mass is zero, so there is no kinetic energy in either rotation or translation. The mass is lifted 1.0 meter, so it has gravitational potential energy of

$$PE = mgh = (0.25)(9.8)(1.0) = 2.45$$

Flywheel

This is the total mechanical energy of the system.

At the end of the run, the potential energy is zero, but both the disk is spinning and the mass is moving. Since there is no loss of mechanical energy assumed, we can write:

$$E_f = E_i$$
$$\Rightarrow (\tfrac{1}{2}I\omega^2 + \tfrac{1}{2}mv^2) = (mgh)$$
$$\Rightarrow (0.005)(\omega)^2 + (0.125)(v)^2 = 2.45$$

We need another equation. Just as the angular acceleration of the disk is related to the linear acceleration of its rim, the angular velocity and linear velocities are related:

$$v = r\omega$$

The linear velocity of the rim must be equal to the linear velocity of the mass because they are connected by a string. We substitute this into the energy equation and get:

$$(0.005)(\omega)^2 + (0.125)(0.10\omega)^2 = 2.45$$
$$\Rightarrow (0.00625)(\omega)^2 = 2.45$$
$$\Rightarrow \omega = 19.8$$

So the final angular speed is 19.8 radians per second. In order to calculate the time involved we need to use the rotational version of one of the kinematic equations in Chapter 2. We know the initial and final angular speeds and we are looking for the time. Thus, we need to know either the angular acceleration and use

$$\theta = \omega_i t + \tfrac{1}{2}\alpha t^2 \qquad (4.4)$$

or the angular displacement and use

$$\Delta\theta = \tfrac{1}{2}(\omega_f + \omega_i)(\Delta t) \qquad (4.5)$$

We do know something about the angular displacement. It is related to the distance the weight falls, which we know. The angular displacement is given by:

$$x = r\theta \Rightarrow (1.0) = (0.10)(\theta) \Rightarrow \theta = 10$$

Thus, the disk rotates 10 radians (which is about 1.6 revolutions). Now we can use Equation (4.5)

$$\theta = \tfrac{1}{2}(\omega_f + \omega_i)(t)$$
$$\Rightarrow (10) = \tfrac{1}{2}((19.8) + (0))(t)$$
$$\Rightarrow t = 0.9899$$

Which is the same as before—as it should be.

Rolling Things

Rolling Things

Suppose we have a sphere, a hollow ball, a ring, and a disk. They all have the same mass (5.0 kilograms) and the same radius (0.2 meters). How long does it take each to roll down a 5.0 meter plane inclined at five degrees?

Answer: Sphere, disk, ball, ring: 4.05, 4.19, 4.42, 4.84 seconds respectively.

The question is the same for the four geometric shapes. Since everything is the same except their moments of inertia, we will leave that as a variable until the end. Well, we can do one better. Each moment of inertia is of the form

$$nmr^2 = (n)(5.0)(0.2)^2 = (0.2)(n)$$

where $n \leq 1$.

The easiest way to solve this is with energy. At the top of the incline each object has no kinetic energy, only potential energy. Since they all have the same mass, they all have the same potential energy, mgh. The height is given by

$$h = (5.0)(\sin 5°) = 0.436$$

Rolling Things

Therefore, the initial potential energy of each is

$$PE = mgh = (5.0)(9.8)(0.436) = 21.35$$

At the bottom of the incline, the objects have *both* types of kinetic energy: $KE = \frac{1}{2}mv^2$ (translation) and $KE = \frac{1}{2}I\omega^2$ (rotation).

We also have some *implicit data* in the problem. When an object is *rolling* (without slipping or sliding), the angular speed and the linear speed must be connected by the equation $v = r\omega$.

Since no energy is lost, we can write:

$$21.35 = \tfrac{1}{2}m(r\omega)^2 + \tfrac{1}{2}(nmr^2)\omega^2$$
$$\Rightarrow 21.35 = (0.1)(\omega)^2 + (0.1)(n)\omega^2$$
$$\Rightarrow \omega = 14.61\sqrt{\frac{1}{1+n}}$$

So the final velocity is

$$v = r\omega = (2.922)\sqrt{\frac{1}{1+n}}$$

Using the kinematic equations in Chapter 2 we have:

$$x = \tfrac{1}{2}(v_i + v_f)(t)$$
$$\Rightarrow (5.0) = \tfrac{1}{2}\left(0 + (2.922)\sqrt{\frac{1}{1+n}}\right)(t)$$
$$\Rightarrow t = 3.422\sqrt{1+n}$$

Therefore...

Object	n	t
Sphere	0.40	4.05
Disk	0.50	4.19
Hollow ball	0.67	4.42
Ring	1.00	4.84

The sphere has the most mass concentrated at the center, so it needn't spend as much energy in rotation and gets to the bottom faster.

This is the last problem dealing with rigid bodies. I have not included any problems dealing with angular momentum because they are usually exact parallels to the previous linear momentum problems in Chapter 3. The subject of rigid body motion can get crazy when you talk about free rotation, gyroscopes, dynamic balancing, etc. But this is usually where the introductory physics class stops, so I will too.

Stretching A Cable

The next subject is elasticity. Truly, there is no such thing as an ideal rigid object. Every solid will bend under sufficient pressure. At every point within the solid there is a complex system of forces in play. The *stress* at each point is described by three components:

Normal stress The force perpendicular to the surface. Either tension or compression.

Shear stress The force parallel to the surface. Causes deflection.

Bulk pressure The pressure all around the object.

Each of these stress components causes a slight deformation called *strain*. The stress and strain are related by an empirical formula. The proportionality constants for the three components of stress are Young's modulus, the shear modulus, and the bulk modulus, respectively.

All three of the components of stress have the same type of formula: the fractional change in length (or volume) is propor-

Stretching A Cable

tional to the stress applied. For tension/compression we have:

$$\frac{F}{A} = Y \left(\frac{\Delta L}{L_0}\right) \tag{4.6}$$

where Y is Young's modulus, the proportionality constant for tension.

For shear stress we have:

$$\frac{F}{A} = S \left(\frac{\Delta x}{L_0}\right) \tag{4.7}$$

But now we have S which is the shear modulus and Δx which is the displacement in the direction of the force perpendicular to the distance L_0.

Finally, for the bulk pressure stress:

$$\Delta P = -B \left(\frac{\Delta V}{V_0}\right) \tag{4.8}$$

We have B which is the bulk modulus and ΔV which is the change in volume V. The minus sign is there because an increase in pressure *decreases* volume. Until now, we have been using F/A rather than pressure directly. The bulk modulus is usually useful in the context of a fluid, which has an internal pressure. Really, pressure is the more "empirical" concept and force the more "ideal" since any real force is applied over a particular area.

For example:

> **Stretching a Cable**
>
> A copper cable is 3.0 meters long with a diameter of 0.050 meters. It supports a 100 kilogram mass. How much is the cable elongated by its support? Young's modulus = 2.0×10^{11}.
>
> **Answer**: 7.5×10^{-6} meters

In our case, the stretching force is the weight of the mass

$$W = mg = (100)(9.8) = 980$$

And the cross-sectional area of the wire is

$$A = \pi r^2 = \pi (0.050/2)^2 = 0.00196$$

Plugging these into Equation (4.6) we have:

$$\frac{(980)}{(0.00196)} = (2.0 \times 10^{11})\frac{(\Delta L)}{(3.0)} \Rightarrow \Delta L = 7.500 \times 10^{-6}$$

> **Shear Stress On A Nail**
>
> A nail with diameter 0.020 meters sticks out from a wall 0.10 meters. How much force is required to bend the nail 1.0×10^{-4} meters? Shear modulus = 3.0×10^{11}.
>
> **Answer**: 95000 newtons

The equation to use is Equation (4.7). In our case,

$$A = \pi r^2 = \pi (0.020/2)^2 = 3.1416 \times 10^{-4}$$

Thus,

$$\frac{(F)}{(3.1416 \times 10^{-4})} = (3.0 \times 10^{11})\frac{(1.0 \times 10^{-4})}{(0.10)}$$
$$\Rightarrow F = 95493$$

Floating Things

> **Volume Compression**
>
> A styrofoam cube 0.10 meters on a side is taken from air pressure to the depths of the ocean in a submarine. The pressure differential is 500 kilopascals. The cube is crushed by the extra pressure to 0.020 meters on a side. What is the bulk modulus of the material?
>
> **Answer:** 5.0×10^{-5} N/m^2

In our case,
$$V = r^3 = (0.10)^3 = 1.0 \times 10^{-3}$$
What is ΔV? It is the *change* in volume, so
$$\Delta V = V_f - V_i = (0.020)^3 - (0.10)^3 = -9.92 \times 10^{-4}$$
Therefore,
$$(5.0 \times 10^5) = -(B)\frac{(-9.92 \times 10^{-4})}{(1.0 \times 10^{-3})} \Rightarrow B = 5.0403 \times 10^{-5}$$

This is very large—usually these moduli are on the order of 10^{-11}. But this is styrofoam so we needn't worry, it *should* be larger than normal.

Often these stress/strain problems are combined with other concepts—using force from the previous chapters, or using temperature which we will cover in the next chapter.

Floating Things

Now we move on to fluids. A fluid is either at rest or flowing. If it is at rest, the basic equation to know is:
$$P_2 - P_1 = \rho g(h_2 - h_1) \tag{4.9}$$

Hydraulic Press

Consider a hydraulic press supporting a car 2.0 meters in the air. The weight of the car is 10,000 newtons and the cross-sectional area of the support is 2.0 m². The cross-sectional support of the press is 5.0 cm². What is the force required at the press? (The density of the fluid in the press is 1.0 g/cm³).

Answer: 12 newtons

There are two principles in play in these hydrostatic problems. The first is the variation of pressure with depth summarized in Equation (4.9). The second is implied by the first: if the depth is the same, the pressure is the same, no matter what the shape of the container. The reason for this is that if there were a pressure differential, the fluid would flow. And the fluid would flow until the pressure differentials were gone. In fact, this is one way to *define* a fluid.

So, the pressure on the press end must be equal to the pressure on the support end when they are at the same height.[4] But the car is higher by two meters. The press end must support the weight of the car *and* the weight of the extra fluid. The mass of the extra weight is given by $m = \rho V$, where ρ is the density of the fluid. In our case, the density is given in non-SI units. We should convert first:

$$\rho = (1.0 \text{ g/cm}^3) \times \left(\frac{1 \text{ kg}}{1000 \text{ g}}\right) \times \left(\frac{100 \text{ cm}}{\text{m}}\right)^3$$
$$= 1000 \text{ kg/m}^3$$

Pay particular attention to the way I have converted the cubic centimeters. A common mistake is to forget to cube the conversion factor. While we are converting, let's change the press

---

[4] This is known as Pascal's Principle.

Floating Things

cross-sectional area:

$$A = (5.0 \text{ cm}^2)\left(\frac{\text{m}}{100 \text{ cm}}\right)^2 = 5.0 \times 10^{-4} \text{ m}^2$$

Now we have all the data in SI units. Let's calculate the total weight of the two meter column of fluid.

$$W = mg = (\rho V)(g) = (\rho)(A)(h)(g)$$
$$= (1000)(2.0)(2.0)(9.8) = 39200$$

Thus, the support at the height of the press end must support a total weight of $10000 + 39200 = 49200$. The pressure at this height is this total force divided by the area over which it is distributed,

$$P = \frac{F}{A} = \frac{(49000)}{(2)} = 24500$$

Since the pressure is the same on the press side, we can calculate the force required:

$$P = \frac{F}{A} \Rightarrow (24500) = \frac{F}{(5.0 \times 10^{-4})} \Rightarrow F = 12.25$$

The real constraint is whether the structure of the hydraulic press can withstand such pressures.

Floating Raft

A raft with a cross-sectional area of 50 m² has a weight of 30,000 newtons. How many people can it support if they each have a weight of 750 newtons? The height of the raft is 0.10 meters.

Answer: 25 people

Archimedes' principle states that the buoyant force in a fluid is equal to the weight of the fluid displaced. This follows from the basic hydrostatic equation above, Equation (4.9). We won't go into the proof now. We will simply use the principle. The maximum buoyant force we should expect from the raft is when it is almost completely submerged. Since we are not told, we will assume the density of the water is 1000 kg/m^3. The mass of the displaced water is this density times the volume of the raft, so the weight is:

$$W = mg = (\rho V)(g) = (\rho)(A)(h)(g)$$
$$= (1000)(50)(0.10)(9.8) = 49000$$

This is the maximum weight the raft can support. Of course it must support its own weight (30,000 newtons), so only 19,000 newtons are left for the people. Dividing by 750 newtons yields 25.3, so the raft can support 25 people.

Flowing Things

There are many ways a fluid can flow—it's really quite complex. The simplest kind of flow is steady, incompressible, ideal fluid flow. For this kind of flow, Bernoulli's Equation holds:

$$P_2 + \rho g h_2 + \tfrac{1}{2}\rho v_2^2 = P_1 + \rho g h_1 + \tfrac{1}{2}\rho v_1^2 \qquad (4.10)$$

This is basically the conservation of energy applied to a fluid. Notice that if the heights are the same, a faster fluid has lower pressure than a slower fluid. This is the principle behind the lift on an airplane's wing and the carburetor in your car.

Flowing Things

> **Open Fountain**
>
> Consider a fountain open to air. The water rises to a height of 10 meters. It comes out of a pipe in the ground with a cross-sectional area of 0.10 m². What is the volume flow rate of this fountain?
>
> **Answer:** 1.4 m³/s

What is this "volume flow rate" mentioned in this problem? It is the number of cubic meters of fluid that flows past per second. We have $Q = Av$, where Q is the volume flow rate, A is the cross-sectional area and v is the speed of the fluid. After we have calculated the speed of the fluid from the base of the fountain we will be able to answer the question using Bernoulli's Equation.

The first thing to note is that the whole situation is exposed to air. Therefore the pressure is atmospheric. But more importantly, $P_1 = P_2$, so these cancel from both sides of the equation. Once we do this, notice there is a factor of ρ in each remaining term, so these cancel. We are left with:

$$gh_2 + \tfrac{1}{2}v_2^2 = gh_1 + \tfrac{1}{2}v_1^2$$

Interestingly, this is the same equation we get when we calculate the trajectory of a projectile using energy. In our case since $v_2 = 0$ (the top of the trajectory, remember), we can solve for v_1:

$$(9.8)(10) + \tfrac{1}{2}(0)^2 = (9.8)(0) + \tfrac{1}{2}(v_1)^2 \Rightarrow v_1 = 14$$

Finally, the volume flow rate is:

$$Q = Av = (0.10)(14) = 1.4$$

Fluid Pressure In A Pipe

A pipe of diameter 0.050 meters has water flowing at 3.0 m/s through it with a volume flow rate of 0.5 m³/s. The pipe rises 2.0 meters and the diameter is constricted to 0.020 meters. What is the difference in pressure between these two points?

Answer: 33 kilopascals

In this case we will be using Bernoulli's Equation in its full power:

$$(P_2 - P_1) = \rho g(h_1 - h_2) + \tfrac{1}{2}\rho(v_1^2 - v_2^2)$$

I've rewritten it in this form since it answers the question directly. Since this is water we can assume $\rho = 1000$ kg/m³. The height differential is 2.0 meters. The speeds we will get from the volume flow rates. But we are only given one number. This is because for an incompressible fluid (which we assume unless otherwise stated), the volume flow rate is the same everywhere. This is simply a statement that the fluid can't "pile up" in the pipe. What goes in must come out. So, the volume flow rate for this problem is:

$$\begin{aligned} Q &= (A_1)(v_1) \\ &= (\pi r_1^2)(v_1) \\ &= (\pi)(0.050/2)^2(3.0) \\ &= 5.89 \times 10^{-3} \end{aligned}$$

This allows us to determine the speed of the fluid at the top of the pipe:

$$\begin{aligned} Q = (A_2)(v_2) &= (\pi r_2^2)(v_2) \\ \Rightarrow (5.89 \times 10^{-3}) &= (\pi)(0.020/2)^2(v_2) \\ \Rightarrow v_2 &= 18.75 \end{aligned}$$

Now we can use Bernoulli's equation:

$$\Delta P = \rho g(h_1 - h_2) + \tfrac{1}{2}\rho(v_1^2 - v_2^2)$$
$$= (1000)(9.8)(2.0) + \tfrac{1}{2}(1000)((5.0)^2 - (18.75)^2)$$
$$= 32719$$

This is about a third of atmospheric pressure, so this seems reasonable.

Tuning A Guitar

So far, we have covered the basic dynamics of "real" solids and fluids. Now we are prepared to talk about waves, a particular type of motion that occurs in both solids and fluids. Sound is a wave transmitted through both solids and fluids. Sound really occurs in three dimensions, but just as we began our discussion of motion in one dimension, it makes sense to talk first about waves in one dimension.

A string is the typical application of one-dimensional wave equations. In fact, one can think about a string as a bunch of oscillators connected together in space. The oscillator picture brings to mind the idea of frequency: the number of cycles per second. But with a string we can also picture a *wavelength*. Frequency has to do with time, wavelength has to do with space. These two ideas are connected by the following simple formula:

$$v = f\lambda \qquad (4.11)$$

Where v is the speed with which a disturbance in the string propagates through space. This in turn is connected to the tension in the string. The greater the tension, the greater the speed. The formula is

$$v = \sqrt{T/\mu} \qquad (4.12)$$

where μ is the linear density of the string. That is the mass of the string divided by its length.

> **Tuning A Guitar**
>
> A guitar string has a linear density of 0.010 kg/m. The C string (512 Hz) is 1.50 meters long. What is the tension in the string?
>
> **Answer**: 24000 newtons

We will assume that the wavelength of the string is twice the length of the string. This is because the wavelength must contain a whole cycle which has a node in the middle. But the string can support half a cycle since it only needs to have a node at each end (none in the middle). Thus, $\lambda = 3.0$. Using Equation (4.11) we have

$$v = f\lambda = (512)(3.0) = 1536$$

Using Equation (4.12) we can determine the tension in the string.

$$v = \sqrt{T/\mu} \Rightarrow (1536) = \sqrt{(T)/(0.010)} \Rightarrow T = 23593$$

> **Pipe Open At One End**
>
> The speed of sound is 340 m/s. A pipe open at one end resonates strongly with a tuning fork tuned to 512 Hz. What are the three smallest lengths the pipe can be?
>
> **Answer**: 0.17, 0.50, and 0.83 meters

The equation to use is

$$f_n = (2n-1)\left(\frac{v}{4L}\right)$$

where $n = 1,2,3,...$

This really follows from Equation (4.11) and the geometry of the pipe. The open end must sit on an anti-node, so the smallest length supports one-quarter a wavelength. Thus, $L = \lambda/4$ and $f = v/4L$. The next length will support three-quarters a wavelength, etc. In our case we have:

$$n = 1 : (512) = (1)(340)/(4L_1) \Rightarrow L_1 = 0.1660$$
$$n = 2 : (512) = (3)(340)/(4L_2) \Rightarrow L_2 = 0.4980$$
$$n = 3 : (512) = (5)(340)/(4L_3) \Rightarrow L_3 = 0.8301$$

Intensity Of Sound

Waves transmit energy (work done here will do work over there). This is true for sound also. *Intensity* measures the power (energy per second) that flows through a unit surface area. If the power source is a point, the conservation of energy implies that the power through any spherical shell must be the same. But since the surface area over which this power is spread is larger, the intensity is smaller. Thus,

$$I = \frac{P}{A} = \frac{P}{4\pi r^2}$$

Here is a problem that uses these ideas.

Sound Intensity

The threshold of hearing is a sound intensity of 1.0×10^{-12} W/m². How powerful a speaker is required for a person to hear at this level 1.0 km away?

Answer: 1.3×10^{-5} watts

For our speaker we have:

$$I = \frac{P}{4\pi r^2}$$
$$\Rightarrow (1.0 \times 10^{-12}) = \frac{P}{(4\pi)(1000)^2}$$
$$\Rightarrow P = 1.2566 \times 10^{-5}$$

Now, this is actually quite small. A real speaker requires more power. Most of the power goes into operating the mechanical parts. But this question also is too simplisitic. True sound intensity falls faster than the geometry suggests due to temperature, air movement, viscosity, etc.

The *decibel* scale is just a different way of measuring intensity. Instead of W/m^2, we use dB. The relationship between the two is:

$$\beta = (10 \text{ dB}) \log_{10}(I/I_0) \qquad \text{or} \qquad I = (I_0) 10^{(\beta/10)} \quad (4.13)$$

where I_0 is the threshold of hearing: 1.0×10^{-12} W/m^2. Why a different scale for the intensity of sound? Because the human ear does not respond linearly with sound intensity. When the intensity doubles, it doesn't *sound* like it has doubled—it sounds much less. The ear (and all the senses, actually) respond much closer to the logarithmic function with which the decibel is defined. Usually, "intensity level" refers to intensity measured in dB, while simple "intensity" refers to intensity measured in SI units (W/m^2).

Using Decibels

A whisper is about 20 dB. What is the relative increase in power required to double the intensity level?

Answer: 100

The "relative increase" is asking for the *ratio* of the final and initial intensities, i.e., I_2/I_1. We can calculate the intensity I_1

using Equation (4.13):

$$I_1 = (I_0)10^{(\beta_1/10)}$$
$$= (1.0 \times 10^{-12})10^{((20)/10)}$$
$$= 1.0 \times 10^{-10}$$

Since the intensity level doubles, we have $\beta_2 = 40$. Therefore the intensity at this level is:

$$I_2 = (I_0)10^{(\beta_1/10)}$$
$$= (1.0 \times 10^{-12})10^{((40)/10)}$$
$$= 1.0 \times 10^{-8}$$

So, the answer is:

$$I_2/I_1 = (1.0 \times 10^{-8})/(1.0 \times 10^{-10}) = 100$$

Every time the intensity level increases by 10 decibels[5] the power must increase ten-fold.

Wave Interference

The quintessential phenomenon associated with waves is interference. Waves can transmit energy, but so do particles. Even diffraction can be understood as the scattering of particles along edges. But particles cannot interfere, waves do. Waves have an additional degree of freedom called *phase*. This is essentially the "location" of the wave within its oscillation. When one wave is "up" and another wave is "down" the two cancel one another (a.k.a., destructive interference). If they are both "up", the phases are aligned and the waves add together (a.k.a., constructive interference).

---

[5] I suppose that is one "bel"...

Wave Interference – Sound

Two speakers are separated by 3.0 meters and emit a pure tone with a frequency of 90 Hz. A microphone is located 4.0 meters directly in front of one of the speakers (forming a 3-4-5 triangle). What phase shift is required between the two speakers to create destructive interference at the microphone? You may assume the speed of sound is 340 m/s.

Answer: 290°

To begin, let's suppose the speakers are in phase (no phase shift). The difference in path-lengths between the speakers and the microphone is 1.0 meter. This is how far the wave from the farther speaker lags behind the nearer one. The wavelength for this sound is given by Equation (4.11):

$$\lambda = v/f = (340)/(100) = 3.4$$

This means the farther speaker is behind by 1.0 / 3.4 = 0.294 of a wavelength. For constructive interference to occur, the farther speaker would have to lead the other by 0.294 wavelengths. But we want destructive interference so we should add another 1/2 of a wavelength (we could subtract also). Thus, we need a phase shift of 0.794 wavelengths. Since 360° corresponds to one wavelength, the phase shift implied is:

$$(0.794 \text{ wavelength}) \times \left(\frac{360°}{1 \text{ wavelength}}\right) \Rightarrow \phi = 290°$$

CHAPTER

FIVE

HEAT AND LIGHT

It is not obvious that either heat or light can be explained using the mechanical principles from the previous chapter. Various models of matter have been tried in the past to explain these two phenomena: fluids (caloric), elastic solids (ether), particles (Newton's "corpuscles" of light).

Today, the model used for heat is the random kinetic energy of molecules and the model used for light is an electromagnetic wave. However, this success with classical mechanics is qualified—that qualification is quantum mechanics. But we can get pretty far using these classical models. That's what we will do in this chapter.

Thermal Expansion

The first thing to know about heat is that it causes expansion. Since expansion is measurable, this gives us a way to *quantify* the sensation of heat. Temperature also has other measurable effects (causes electric current in certain combinations of materials, for example). Collectively these are called *thermometric effects*. Any one of these can be used to quantify temperature, but thermal expansion is the easiest. The formula for thermal expansion is the same as the deformation due to stress:

$$\Delta L/L_0 = \alpha \Delta T \tag{5.1}$$

where α is called the coefficient of *thermal expansion*. These problems are very similar to the stress/strain problems in the previous chapter.

Thermal Expansion – Concrete

You notice that the concrete slabs (0.50 meters long) in your sidewalk are separated by 2.0 millimeters. If $\alpha = 2.0 \times 10^{-5}$, how much can the temperature increase before the slabs touch?

Answer: 200 °C.

Remember that all the slabs expand, so the two slabs forming the separation are expanding toward each other. They each need to cover a gap of 1.0 millimeter. On the other hand, the slabs expand from the center, so the length $L_0 = 0.25$ meters. Thus,

$$(1.0 \times 10^{-3})/(0.25) = (2.0 \times 10^{-5})(\Delta T) \Rightarrow \Delta T = 200$$

Thermal Expansion – Fracturing

Suppose a certain species of concrete ($\alpha = 1.0 \times 10^{-5}$, $Y = 2.5 \times 10^{10}$) will begin to show fractures at a compressive stress of 20,000 kPa. If a concrete slab 1.0 meter in length is constrained into this length, what change in temperature is required to produce fractures?

Answer: 80 °C

Thermal Expansion   117

An increase in temperature will cause the concrete to expand. But the constraint force from the wall responds with a compressive force to push the concrete back to the original length. This compressive force is governed by the stress-strain relationship from the previous chapter:

$$F/A = Y\frac{\Delta L}{L_0}$$

$$\Rightarrow (2.0 \times 10^7) = (2.5 \times 10^{10})\frac{\Delta L}{(1.0)}$$

$$\Rightarrow \Delta L = 8.0 \times 10^{-4}$$

What temperature differential will produce this expansion?

$$(8.0 \times 10^{-4})/(1.0) = (1.0 \times 10^{-5})(\Delta T) \Rightarrow \Delta T = 80$$

So, don't go building any bridges with this stuff.

These same ideas apply to volume as well. The relevant equation is

$$\Delta V/V_0 = \beta \Delta T$$

where β is the thermal coefficient of volume expansion.

Thermal Expansion – Volume

Suppose a glass flask that holds one liter (1.0×10^{-3} m^3) is filled to the brim with water. If the temperature is raised 50 °C, how much water spills out? The coefficients of volume expansion for the glass and water are 2.0×10^{-4} and 3.0×10^{-4}, respectively.

Answer: 5.0×10^{-6} m^3

Both the glass and the water expand. First the water:

$$(\Delta V)/(1.0 \times 10^{-3}) = (3.0 \times 10^{-4})(50)$$

$$\Rightarrow \Delta V = 1.5 \times 10^{-5}$$

Now the glass:

$$(\Delta V)/(1.0 \times 10^{-3}) = (2.0 \times 10^{-4})(50)$$
$$\Rightarrow \Delta V = 1.0 \times 10^{-5}$$

You might object at this point by saying that since the glass expands the interior must shrink. Actually, this is not the case. This is most easily seen by imaging a 3×3 grid of nine squares. If they expand, so does the center. All linear dimensions expand including the border of a hole.

So, the glass expands 1.0×10^{-5} m^3, but the water expands more:

$$1.5 \times 10^{-5} - 1.0 \times 10^{-5} = 5.0 \times 10^{-6}$$

Calorimetry

There are really two concepts wrapped up in the ideas of heat and temperature. Physicists associate the first of them with the term *temperature*. This is the physical, measurable manifestation of the phenomena. On the other hand, *heat* is the energy required to produce the physical effect. In a sense, the distinction is one between cause and effect: heat flow causes a change in temperature. The distinction is important because often heat flow does not cause a change in temperature. This occurs when a substance boils or melts, for example.

When a substance absorbs heat by changing temperature, the rate at which it absorbs heat is called its *specific heat*. When a substance absorbs heat by changing phase, the similar term is called its *latent heat*.[1] This leads to the two basic equations

---

[1] Temperature is a measure of the random kinetic energy of molecules. A phase change can be seen as the change in the *potential* energy of molecules.

Calorimetry

relating heat and temperature:

$$\Delta Q = mc\Delta T \qquad (5.2)$$

and

$$\Delta Q = mL \qquad (5.3)$$

where c is the specific heat and L is the latent heat of the substance.

> **Heat Flow – Phase Change**
>
> How much heat is required to take a 1.0 kilogram block of ice from -10 °C to steam at 120 °C?
>
> **Answer:** 3.30×10^6 joules

This is a multi-step process:

- Heat the ice from -10 °C to 0 °C
- Melt the ice
- Heat the water from 0 °C to 100 °C
- Boil the water
- Heat the steam from 100 °C to 120 °C

The specific heat of ice is 2050 J/kg-K, so the heat required to raise the temperature of the ice from -10 °C to 0 °C is

$$\Delta Q = mc\Delta T = (1.0)(2050)(10) = 20500$$

The latent heat (fusion) of ice/water is 3.34×10^5 J/kg, so the heat required to melt the ice is

$$\Delta Q = mL = (1.0)(3.34 \times 10^5) = 3.34 \times 10^5$$

The specific heat of water is 4181 J/kg-K, so the heat required to raise the temperature of the water from 0 °C to 100 °C is

$$\Delta Q = mc\Delta T = (1.0)(4181)(100) = 4.181 \times 10^5$$

The latent heat (vaporization) of water/steam is 2.50×10^6 J/kg, so the heat required to boil the water is

$$\Delta Q = mL = (1.0)(2.50 \times 10^6) = 2.50 \times 10^6$$

The specific heat of steam is 2027 J/kg-K, so the heat required to raise the temperature of the ice from 100 °C to 120 °C is

$$\Delta Q = mc\Delta T = (1.0)(2027)(20) = 40540$$

The total heat required is the sum of all of these steps.

$$\begin{aligned}\Delta Q &= (20500) + (3.34 \times 10^5) + (4.181 \times 10^5) \\ &\quad + (2.50 \times 10^6) + (40540) \\ &= 3.30 \times 10^6\end{aligned}$$

Heat Flow – Final Temperature

If you take 0.10 kilograms of ice at -10 °C and add it to 2.0 kilograms of hot water as 60 °C. What is the final temperature of the mixture?

Answer: 53 °C

This also is a multistep process.

- Heat the ice from -10 °C to 0 °C
- Melt the ice
- Heat the ice-water from 0 °C to the final temp (T_f)

The heat required to perform this process must come from the hot water. Since heat leaves the hot water, it must cool down:

Calorimetry

- Cool the hot-water from 60 °C to the final temp (T_f)

Now, it is possible that there is enough ice to bring the hot water temperature down to its freezing point, but in this problem that seems unlikely. So the next thing to do is to calculate the heat involved in each step we have laid out.

The specific heat of ice is 2050 J/kg-K, so the heat required to raise the temperature of the ice from −10 °C to 0 °C is

$$\Delta Q = mc\Delta T = (0.10)(2.050)(10) = 2050$$

The latent heat (fusion) of ice/water is 3.34×10^5 J/kg, so the heat required to melt the ice is

$$\Delta Q = mL = (0.10)(3.34 \times 10^5) = 33400$$

The specific heat of water is 4181 J/kg-K, so the heat required to raise the temperature of the water from 0 °C to T_f is

$$\Delta Q = mc\Delta T = (0.10)(4181)(T_f) = (418.1)(T_f)$$

So, the total heat going into the ice is

$$\Delta Q = (35450) + (418.1)(T_f)$$

Again, this heat comes from the hot water. Since the specific heat of water is 4181 J/kg-K, the total heat from the hot water is

$$\begin{aligned}\Delta Q &= mc\Delta T \\ &= (2.0)(4181)(60 - T_f) \\ &= (501720) - (8362)(T_f)\end{aligned}$$

These two quantities of heat must be equal. This allows us to solve for the final temperature.

$$(35450) + (418.1)(T_f) = (501720) - (8362)(T_f)$$
$$\Rightarrow T_f = 53$$

Ideal Gas Law

Gases are important idealizations. Technically, a gas is a type of fluid because it can flow. Therefore the only "component" of internal stress it can support is pressure. Unlike liquids, gases expand to fill their container. At a molecular level, this is because the interaction between the molecules is very slight, so they don't "clump" together like a liquid or solid. A gas is an *ideal gas* when this interaction is zero.

So, ideal gases are simple. We can consider them as a collection of independent particles travelling in an enclosed volume. The pressure the gas exerts is from the rebounding particles off of the container (similar to the rain pressure problem in Chapter 4). This kinetic model of gases can be used to derive many properties. The most important is the Ideal Gas Law:

$$PV = nRT = NkT \tag{5.4}$$

where n is the number of moles of gas or N is the number of molecules of gas; R is called the "ideal gas constant" and k is called "Boltzmann's constant".

Notice that pressure multiplied by volume has the units of energy. Therefore, so does the right side of the Ideal Gas Law. The energy combination kT will come up frequently.

This is the first formula we have dealt with so far that deals with temperature directly rather than a *change* in temperature. When calculating change, both the Celsius and *Kelvin* scales give the same answer. But when dealing with the absolute temperature, we *must* use Kelvin.

Ideal Gas Law

> **Ideal Gas – Calculate Pressure**
>
> An ideal gas is contained in a particular volume. If the pressure is 200 kilopascals when the temperature is 30 °C, what is the pressure when the temperature rises to 200 °C?
>
> **Answer**: 310 kilopascals

We are dealing with a problem that starts in one state and ends in another state. It is usually easier to take the basic equation $PV = nRT$ and divide it by itself in the following way:

$$\frac{P_2}{P_1}\frac{V_2}{V_1} = \frac{n_2}{n_1}\frac{R}{R}\frac{T_2}{T_1}$$

Obviously the R cancels, but so does $n_2 = n_1$ and $V_2 = V_1$. This cancellation is precisely why we want to divide this equation with itself. So, we end up with

$$\frac{P_2}{P_1} = \frac{T_2}{T_1}$$

Into which we can simply plug our data:

$$\frac{P_2}{2.0 \times 10^5} = \frac{473}{303} \Rightarrow P_2 = 3.1221 \times 10^5$$

For an ideal gas, the average energy associated with a monatomic molecule is $\frac{3}{2}kT$. Since there is no potential energy for an ideal gas (no interaction), all of this energy is kinetic.

> **Ideal Gas – Average Speed**
>
> A 2.0 m³ container of gas holds 50 moles of argon gas at atmospheric pressure. What is the average speed of a molecule of oxygen gas in this container?
>
> **Answer**: 550 m/s

The average speed of the gas is represented in its temperature. Using the Ideal Gas Law we can determine the temperature of the gas from the data given.

$$PV = nRT$$
$$\Rightarrow (1.0 \times 10^5)(2.0) = (50)(8.31)(T)$$
$$\Rightarrow T = 481$$

Since argon gas is monatomic, the average internal energy per molecule is

$$U = \tfrac{3}{2}kT = \tfrac{3}{2}(1.38 \times 10^{-23})(481) = 9.96 \times 10^{-21}$$

We assume all of this is kinetic energy. The mass of one argon atom in SI units is 6.63×10^{-26} kilograms. From the definition of kinetic energy ($KE = \tfrac{1}{2}mv^2$), we have:

$$(9.96 \times 10^{-21}) = \tfrac{1}{2}(6.63 \times 10^{-26})(v)^2 \Rightarrow v = 548.1$$

Note that this is the average *speed*. The average velocity is zero.

Adiabatic Expansion

Until now we have spoken about heat flow in general. We will now begin to answer questions about harnessing the flow of heat. We seek to build a heat engine: a machine that will convert the flow of heat into useful work. We will discover that even in the best of situations, it is not possible to put all the heat energy to work.

In preparation, we should discuss the various circumstances under which heat flows. There are four "ideal" thermal processes to consider.

- Those at constant volume (isochoric)

Adiabatic Expansion

- Those at constant pressure (isobaric)
- Those at constant temperature (isothermal)
- Those with no heat flow (adiabatic)

Processes which are very fast can be assumed adiabatic: explosions, for example.

The work associated with each of these four types of processes is as follows:

- $W_{\text{isoc}} = 0$
- $W_{\text{isob}} = P\Delta V$
- $W_{\text{isot}} = nRT\ln(V_f/V_i)$
- $W_{\text{adia}} = -\frac{3}{2}nR(\Delta T)$

The first and second follow from the definition of work, the third from the ideal gas law, and the fourth from the internal energy of an ideal gas.

There is one more important equation for the adiabatic process. If we give the relative change in volume (V_f/V_i) a label of x, then from the ideal gas law the relative change in pressure for an isothermal process is $1/x$. For an adiabatic process, the relative change in pressure is $(1/x)^\gamma$, where γ is a number greater than one. When volume increases, the pressure in an adiabatic will decrease *faster* than in an isothermal one. The adiabatic equation is usually represented as follows:

$$P_i V_i^\gamma = P_f V_f^\gamma \qquad (5.5)$$

The quantity γ can be shown to be equal to the ratio of the specific heat of the gas at constant pressure versus the specific heat at constant volume.

> **Ideal Gas – Work Done**
>
> A container with 1.0 moles of an ideal gas (with $\gamma = 7/5$) begins with a volume of 0.020 m³ and a temperature of 300 kelvin. The final volume is 0.010 m³. What is the difference in work required for an isothermal process rather than an adiabatic one?
>
> **Answer:** -533 joules

The work from the isothermal process is:

$$W_{\text{isot}} = nRT \ln\left(\frac{V_f}{V_i}\right)$$
$$= (1.0)(8.31)(300) \ln\left(\frac{(0.010)}{(0.020)}\right)$$
$$= -1728$$

Work must be done to compress the gas at a constant temperature.

In order to calculate the work from the adiabatic process, we need to determine the final temperature. We could use the Ideal Gas Law if we knew the pressure and volume. But how do we get the volume? We use Equation (5.5). Okay, but what is the initial pressure? We can use the Ideal Gas Law for this too.

$$(P_i)(0.020) = (1.0)(8.31)(300) \Rightarrow P_i = 1.25 \times 10^5$$

Now we use the adiabatic PV formula, Equation (5.5):

$$(1.25 \times 10^5)(0.020)^{7/5} = (P_f)(0.010)^{7/5}$$
$$\Rightarrow P_f = 3.29 \times 10^5$$

And we use the Ideal Gas Law again (on the final state, now):

$$(3.29 \times 10^5)(0.010) = (1.0)(8.31)(T_f) \Rightarrow T_f = 396$$

Finally, we use the adiabatic work formula:

$$W_{\text{adia}} = -\tfrac{3}{2}(1.0)(8.31)(396 - 300) = -1195$$

Work must also be done on the gas to maintain the adiabatic process, but less than that for the isothermal process.

Think of it this way. The adiabatic process is what happens when the system is isolated—the temperature naturally rises when the gas is compressed. In order to support the process isothermally, there must be some heat reservoir. The reservoir draws some of the energy (as heat) from the work done in compressing the gas. So, more work needs to be done to get the same volume compression.

The difference in work done is:

$$W_{\text{isot}} - W_{\text{adia}} = (-1728) - (-1195) \Rightarrow \Delta W = -533$$

This is the amount of heat that is pushed into the heat reservoir to maintain the temperature at 300 kelvin.

Carnot Heat Engine

Friction converts mechanical energy into heat. The previous sections show that we can convert heat back into mechanical energy (i.e., work). In previous chapters, we had considered this mechanical energy as lost. But we *can* recover some of this energy. This justifies considering heat as another form of energy. This means we ought to generalize the conservation of energy to include heat. This is known as the First Law of Thermodynamics.

$$\Delta U = \Delta Q - \Delta W \qquad (5.6)$$

where ΔU represents the internal energy of the system, ΔQ is the heat put *into* the system, and ΔW is the work done *by* the system. In words, after taking into account the work the system does, any heat left over increases internal energy.

First Law Of Thermodynamics

A monatomic gas (30 moles) is allowed to expand from 1.0 m^3 to 1.1 m^3 at atmospheric pressure. Due to friction in the system, 5000 joules of heat are added to the gas. What is the final temperature of the gas?

Answer: 380 kelvin

A change in temperature is related to a change in internal energy. According to the First Law, this in turn is related to the work done on the gas and any incremental heat flow. We are given the heat flow, what about work? Since the process is isobaric (constant pressure), the work done on the gas is:

$$W = P\Delta V = (1.0 \times 10^5)(1.1 - 1.0) = 10000$$

This is positive because the gas is doing work. The total change in internal energy is:

$$\Delta U = \Delta Q - \Delta W = (5000) - (10000) = -5000$$

The net change in internal energy is negative, so we should expect a decrease in temperature:

$$\Delta U = nR\Delta T \Rightarrow (-5000) = (30)(8.31)(\Delta T) \Rightarrow \Delta T = -20$$

Without the addition of heat, the temperature difference would be twice this amount.

We are not quite done. We need the "final temperature". What we have is the *change* in temperature. The initial temperature can be determined by using the Ideal Gas Law.

$$PV = nRT$$
$$\Rightarrow (1.0 \times 10^5)(1.0) = (30)(8.31)(T)$$
$$\Rightarrow T_i = 401$$

Carnot Heat Engine

Therefore,

$$T_f = T_i + \Delta T = (401) + (-20) \Rightarrow T_f = 381$$

The first law of thermodynamics expresses the mechanical equivalence of heat. A heat engine takes a temperature differential and converts the heat flow into work. The efficiency of any engine is the ratio of its work output over its energy input. For a heat engine:

$$e = \frac{W}{Q_H} \Rightarrow e = 1 - \frac{Q_C}{Q_H}$$

This follows because $Q_H = W + Q_C$. The energy that goes in must come out.

The obvious question becomes: what is the maximum efficiency at which a heat engine can be built? Carnot answered this question. The answer consists of three parts:

- Carnot imagined a reversible engine. A *reversible process* is one which can be reversed without any loss of energy (or net work done).

- Carnot proved that all reversible engines working between two heat reservoirs have the *same* efficiency. The efficiency of a reversible engine can only depend on the temperature of the reservoirs.

- Any other (irreversible) engine must be less efficient than a reversible one operating between the same heat reservoirs.

Since all reversible engines have the same efficiency, Carnot analyzed the simplest possible one based on an ideal gas. He found that the efficiency was related to the temperature in the following way:

$$e_{\text{rev}} = 1 - \frac{T_C}{T_H} \tag{5.7}$$

This is the maximum possible efficiency for a heat engine. Most real engines operate at 30% to 60% of this ideal efficiency.

Building A Better Heat Engine

A reversible heat engine operates between heat baths at 300 kelvin and 700 kelvin. Which will increase the efficiency more: increasing the hot bath by 100 kelvin or decreasing the cold bath by 100 kelvin?

Answer: One should decrease the cold bath.

The efficiency of the heat engine is

$$e = 1 - \frac{T_C}{T_H} = 1 - \frac{(300)}{(700)} = 0.57$$

If the hot bath is increased to 800 kelvin, the efficiency becomes

$$e = 1 - \frac{T_C}{T_H} = 1 - \frac{(300)}{(800)} = 0.63$$

If the cold bath is decreased to 200 kelvin, the efficiency becomes

$$e = 1 - \frac{T_C}{T_H} = 1 - \frac{(200)}{(700)} = 0.71$$

Obviously the one with the cold bath is better.

This shows that not all degrees of temperature are created equal. The lower temperature ranges are more "powerful" in some sense. This property is captured in the idea of entropy below.

What About Entropy?

> **Efficiency Ratings**
>
> A irreversible heat engine operating between baths at 500 kelvin and 200 kelvin produces 100 watts of power while expelling 200 watts in heat. What is its efficiency rating relative the maximum possible reversible efficiency?
>
> **Answer**: 55%

The actual efficiency is, by definition,

$$e = W/Q_H = (100)/(300) = 0.33$$

where $Q_H = 400$ watts since $Q_H = W + Q_C$. A reversible engine would have an efficiency of

$$e = 1 - T_C/T_H = 1 - (200)/(500) = 0.60$$

That means the real engine is operating at about 55% of the best possible engine.

What About Entropy?

Entropy is a difficult beast to define. So difficult, in fact, that many introductory textbooks and study guides avoid the topic altogether. So if you don't need this section, skip it! On the other hand, the experts assert that a proper understanding of entropy is the heart of thermodynamics. So, it seems appropriate to at least touch the subject here.

Basically, *entropy* is a measure of the distribution of heat energy within a system. The broader the distribution, the larger the entropy. A system with low entropy is one in which the heat is highly concentrated. It is important to remember that a system *has* entropy like the system *has* internal energy. It is a property of the system.

Nearly every mechanical or thermal process changes the entropy of a system. The only exception is a reversible adiabatic process. Unfortunately *how much* the entropy changes is nearly impossible to calculate. For a reversible isothermal process the change in entropy is:

$$\Delta S = \Delta Q / T \qquad (5.8)$$

If the process is reversible but not isothermal, one can apply this in a differential form using some calculus.

So, in principle, we can calculate the entropy change when a system undergoes any reversible process. What about an irreversible process? The answer *must* be the same. The entropy change depends only on the initial and final states of the system, not in the way it gets there.

Seems simple enough. But there is something hidden in this principle...

Calculate Entropy – Free Expansion

One mole of an ideal gas is constrained to a 1.0 m^3 volume. The gas then expands freely (no outside pressure) to a volume twice the original size. The temperature remains 300 kelvin during this free expansion. How much entropy was created by the system?

Answer: 5.8 J/K

Since the amount of entropy added is the same regardless of whether the process is reversible or not, the first step is to imagine a reversible process that moves between the same initial and final states. The temperature stays constant, so we can imagine another system in which the gas expands isothermally against a slow moving piston. Notice how we are forced to add elements to the system: a piston to absorb the work and a heat reservoir to maintain the temperature of the gas.

What About Entropy?

The gas is ideal, so the internal energy is proportional to the temperature. Since the change in temperature is zero, so is the change in internal energy. Therefore the First Law reduces to $\Delta Q = \Delta W$. This means that the work done on the piston is compensated by a flow of heat from the reservoir rather than a decrease in temperature (and internal energy) of the gas.

The work required for an isothermal process is

$$\Delta W = nRT \ln(V_2/V_1) = (1)(8.31)(300)\ln(2) = 1728$$

Thus, $\Delta Q = 1728$. By definition,

$$\Delta S = \Delta Q/T = (1728)/(300) = 5.76$$

The heat flow from the reservoir has increased the entropy of the gas. Of course, the entropy of the reservoir has decreased by the same amount because $\Delta S = \Delta Q/T = (-1728)/(300) = -5.76$.

This entropy trade-off is true of all reversible processes. When both the system and its surroundings are taken into account, a reversible process will conserve entropy. This is the first half of the Second Law of Thermodynamics.

What about the irreversible process? The change in the system's entropy is the same regardless of whether the process is reversible or irreversible. This is why we replaced the system with a reversible one. So, the change in entropy is the same: $\Delta S = +5.8$ J/K. But in our irreversible case, there is no corresponding decrease in entropy from the surroundings. When both the system and its surroundings are taken into account, an irreversible process creates entropy. This is the second half of the Second Law of Thermodynamics.

These two halves are often summarized in the following way:

$$\Delta S_{\text{univ}} \geq 0 \qquad (5.9)$$

The subscript "univ" is supposed to mean "universe". It's a little silly to say it this way. In fact, strictly speaking, it doesn't even make sense to speak of the entropy of the universe as a whole (nor energy either). Just think "system plus surroundings" or "isolated system" instead. Remember that the equality only holds for reversible processes. Since reversible processes are infinitely slow, the inequality holds for any real process.

One more problem:

> **Another Entropy Problem**
>
> Two identical masses are separated by an thermally insulating partition. One is at 300 kelvin and the other at 200 kelvin. The partition is removed and 100 joules of heat flows in one second. Assume the temperature change is negligible. What is the change in entropy of the combined system?
>
> **Answer:** 0.167 J/K

This is an irreversible process. But it is simple to imagine a reversible substitute by incorporating additional elements from the surroundings. Pull the 100 joules from the 300 kelvin mass to a third mass in the surroundings with the same temperature. The change in entropy is:

$$\Delta S_h = \frac{\Delta Q}{T_h} = \frac{(-100)}{300} = -0.3333$$

Similarly, push 100 J from a fourth mass in the surroundings at 200 K into the second mass. The change in entropy here is:

$$\Delta S_c = \frac{\Delta Q}{T_c} = \frac{(+100)}{200} = +0.5000$$

Therefore, the total change in entropy in the *system* is $\Delta S_{\text{tot}} = +0.1667$.

Since this process is irreversible, no usable work was extracted. The maximum amount of work that could have been extracted is with a reversible engine operating between the two temperatures. If the engine is reversible then the entropy is conserved. Therefore, the entropy extracted from the hot reservoir is equal to the entropy created in the cold reservoir.

However, the entropy created in the *coldest* reservoir of the system cannot ever be extracted because heat will not flow out of it. This heat is *lost* in terms of usable work. In general, the energy that becomes unavailable for work due to an increase in entropy is

$$Q_{\text{lost}} = T_0 \Delta S$$

Where T_0 is the coldest reservoir in the system.

Light On Glass

Now for a complete change of subject. Let's talk optics. In optics we ignore the electromagnetic origin of light and take light on its own terms. The discipline spits into two main pieces: geometric optics and physical optics.

1. **Geometric Optics**

    The study of light which ignores its wave characteristics. This is usually okay because the wavelength of the light wave is so small (350 to 650 nm). Therefore we speak of *light rays*, which bring to mind the image of little particles of light flying around. Be careful not to push this idea too hard, because it won't work (Newton tried). But the geometric notion of light rays is useful nonetheless.

2. **Physical Optics**

    The study of the wave nature of light. This includes the topics of diffraction and interference we talked about in Chapter 4.

Chapter 5. Heat and Light

The simplest problem in geometric optics is reflection. The basic rule is that the reflected angle is the same as the incident angle.

Reflection Targeting

A boy is playing with a laser. He is hoping to hit a target via a mirror located 1.0 meter to his right. The target is 1.0 meter to his left and 3.0 meters straight ahead. At what angle should he aim the laser to hit his target?

Answer: 45°

At this point all we really know is that the reflected angle must equal the incident angle. It may seem like we need to check all of the possible bounces and pick the one which has this characteristic. But this is the wrong way to approach these problems. The key is in a new notion, that of an optical image.

An *image* is the point in space from which the light rays *appear* to emanate. In this case, the image is located in the following way: start from the target and draw a perpendicular line to the mirror. Continue the line through the mirror for the same distance. The image is at this point. The geometry is such that every line originating from this point on the other side of the mirror is where a true light ray would bounce toward the target.

So the problem is really about finding the image. Where is it? Well, how far from the mirror is the target? The target is "1.0 meter to his left" and the mirror is "1.0 meter to his right". The total distance is 2.0 meters. Therefore the image is 2.0 meters to the right side of the mirror. This is 3.0 meters to the boy's right.

The boy needs to aim at the image in the mirror. The image is located 3.0 meters ahead and 3.0 meters to the right. That

Light On Glass

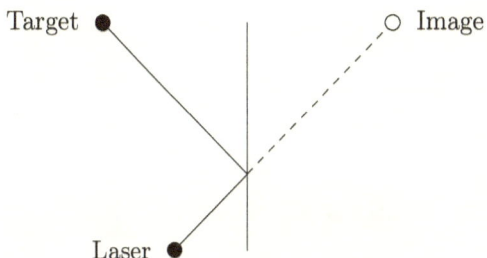

Figure 5.1: Finding the mirror image

is an angle of 45°.

When light strikes a transparent surface, it is both partially reflected and refracted. That is, some of the light bounces and some passes through. The rays that pass through the glass are bent at an angle. This rule is called Snell's Law:

$$n_1 \sin \theta_1 = n_2 \sin \theta_2 \qquad (5.10)$$

The angles in this formula are measured from a line perpendicular to the surface: a zero degree angle is straight into the surface. The symbol n is called the *index of refraction* and is a characteristic of the material. Glass is usually about 1.5, diamond is 2.4 (this is large), while air and vacuum are both 1.0.

The index of refraction is related to the speed of light. The higher the index of refraction, the slower the speed. In fact,

$$v = c/n \qquad (5.11)$$

where c is the speed of light in a vacuum. At an interface of two different indices of refraction, the speed changes. The incident and refracted waves are in sync at the interface: when one is up the other is up, etc. This means that the two waves have the same frequency. Therefore, the wavelengths must be different because all three are related according to $v = f\lambda$ from

138 Chapter 5. Heat and Light

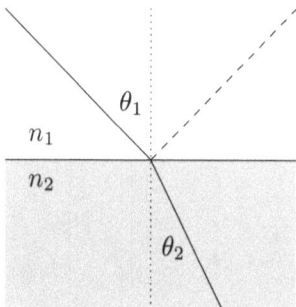

Figure 5.2: Snell's law of refraction

Chapter 4. This implies:

$$n_2 \lambda_2 = n_1 \lambda_1$$

You can see how we are halfway to deriving Snell's Law.

Dispersion Through A Prism

A white light beam strikes a 60-60-60 prism parallel to its base. What is the angular spread of the out-going multi-color beam? The index of refraction for red and blue are 1.54 and 1.50 respectively.

Answer: 2.99°

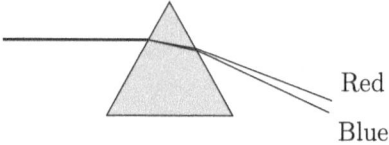

Figure 5.3: Making a rainbow

The light beam is moving horizontally and strikes the first surface. The angle formed with the perpendicular normal line is

30°. This is θ_1. Since the medium of the incident ray is the air, $n_1 = 1.0$. The refracted medium is the glass so $n_2 = 1.50$. Snell's law tells us the internal refracted angle:

$$(1.00)\sin(30°) = (1.50)\sin(\theta_2) \Rightarrow \theta_2 = 19.47°$$

Here is a good rule of thumb: the refracted ray bends *into* the higher index of refraction. So the ray travels through the peak of the prism now aimed slightly down: an angle of $-10.53°$. When it strikes the other side, the angle of incidence is $40.53°$ ... this is the new θ_1. The incident medium is the glass so $n_1 = 1.50$. Appling Snell's Law again yields:

$$(1.50)\sin(40.53°) = (1.00)\sin(\theta_2) \Rightarrow \theta_2 = 77.10°$$

This is an angle of $-47.10°$ from the original beam.

Now for the blue beam. The logic is the same...

$$(1.00)\sin(30°) = (1.54)\sin(\theta_2) \Rightarrow \theta_2 = 18.95°$$

which is $-11.05°$ from the original beam. The angle of incidence on the second surface is $41.05°$, so

$$(1.50)\sin(41.05°) = (1.00)\sin(\theta_2) \Rightarrow \theta_2 = 80.09°$$

This is an angle of $-50.09°$ from the original beam.

Therefore the angular spread is $(-50.09) - (-47.10) = 2.99°$.

Lenses

So we can control light rays by bouncing and bending them. This allows the construction of lenses. A lens is an optical system that focuses light to a particular point in space. By focusing light, the lens creates an image of the object. The *focal point* of a lens is the location of an image when the object

Type	Object	Image	Focal Length
Mirror	To Left	To Left	Concave
Lenses	To Left	To Right	Convex

Table 5.1: Lens Sign Convention

is infinitely far away. In other words, when the incoming light rays are parallel to the axis of the lens.

The image, object and focal length distances obey the following relationship:

$$\frac{1}{d_o} + \frac{1}{d_i} = \frac{1}{f} \tag{5.12}$$

In this equation it is important to maintain a proper sign convention. Table 5.1 summarizes when the length is considered positive.

In addition, the magnification (either mirror or lens) is given by:

$$m = \frac{h_i}{h_o} = -\frac{d_i}{d_o} \tag{5.13}$$

Here the sign convention is maintained. If the magnification is negative, the image is "inverted" (or upside-down).

Concave Mirror

An object sits 5.0 meters in front of a concave mirror with a focal length of 1.0 meter. Where is the image and what is its magnification?

Answer: The image is at 1.25 meters in front of the mirror, is reduced to 25% of the original size, and is inverted

This problem is straightforward. The only trick is to apply the sign convention correctly. The image distance d_i is given by Equation (5.12).

$$\frac{1}{(5.0)} + \frac{1}{(d_i)} = \frac{1}{(1.0)} \Rightarrow d_i = 1.2500$$

This is 1.25 meters in front of the mirror since the answer is positive. And the magnification is given by Equation (5.13).

$$m = -\frac{(1.25)}{(5.0)} = -0.2500$$

Convex Mirror

An object sits 5.0 meters in front of a convex mirror with a focal length of 1.0 meter. Where is the image and what is its magnification?

Answer: The image is at 0.83 meters behind the mirror, is reduced to 14% of its original size, and upright

Similar problem. The image distance d_i is:

$$\frac{1}{(5.0)} + \frac{1}{(d_i)} = \frac{1}{(-1.0)} \Rightarrow d_i = -0.8333$$

So, this is 0.83 meters *behind* the mirror. Convex mirror always have images behind the mirror. (Can you show this?) The magnification is

$$m = -\frac{(-0.83)}{(6.0)} = 0.1383$$

Optical Systems

Now for a more difficult optics problem. An optical system is a combination of mirrors and lenses. Microscopes and telescopes are basic optical systems.

Double Lenses – Calculate Image Distance

Two lenses are separated by 4.0 centimeters each with a focal length of 2.0 centimeters. An object 10.0 centimeters in front of the first lens. How far away is the image from the second lens?

Answer: 4.5 centimeters

In this problem we start by ignoring the second lens. The image through the first lens is given using Equation (5.12):

$$\frac{1}{(10.0)} + \frac{1}{(d_i)} = \frac{1}{(2.0)} \Rightarrow d_i = 0.40$$

Remember that the image is the point in space in which the light rays appear to be generated. Thus, the image of the first lens is the object of the second lens.[2]

Since the image is 0.40 centimeters to the right of the first lens, it is 3.60 centimeters in front of the second lens. This is the object distance for the second lens. Its image is located at

$$\frac{1}{(3.60)} + \frac{1}{(d_i)} = \frac{1}{(2.0)} \Rightarrow d_i = 4.500$$

This is the distance to the right of the second lens.

---

[2] This is always true. The weirdest case is when the object distance happens to be negative (object on the wrong side). The geometry is still captured in Equations (5.12) and (5.13).

Optical Systems

> **Double Lenses – Calculate Focal Length**
>
> Two lenses are separated by 2.0 centimeters each. The first lens has a focal length of 4.0 centimeters. An object is 10.0 centimeters in front of the first lens casts an image 5.0 centimeters to the right of the second lens. What is the focal length of the second lens?
>
> **Answer:** 16 centimeters

Again, the image from the first lens is

$$\frac{1}{(10.0)} + \frac{1}{(d_i)} = \frac{1}{(4.0)} \Rightarrow d_i = 6.67$$

Since this is larger than the distance separating the lenses, the object distance in the second lens is *negative*. It is -4.67 centimeters. Since the image is 5.0 centimeters, we can determine the focal length via Equation (5.12).

$$\frac{1}{(3.60)} + \frac{1}{(-4.67)} = \frac{1}{f} \Rightarrow f = 15.712$$

In optometry, the lens strength is often quoted in *diopters*. This is called the refractive power of the lens, which is nothing more than the reciprocal of the focal length. This is useful for two reasons. First, the larger the refractive power, the stronger the lens. So, stronger lenses have bigger numbers. Second, for thin lenses in contact, the combined refractive power is the sum of the individual refractive powers. Compare with Equation (5.12).

> **Near-Sighted Prescription**
>
> A near-sighted person has a far-point of 500 centimeters. He wears glasses that are 2.0 centimeters in front of his eye. What is the proper prescription for his glasses?
>
> **Answer:** 0.20 diopters

The far-point is the maximum distance the eye can focus an image on its retina. The eye can deal with any distance nearer (hence "near-sighted"). Glasses need to take an object that is infinitely far away ($d_o = \infty$) and cast it to an image on the far point. The lens in the eye will then be able to focus the image into the retina of the eye with its maximum focusing strength.

$$\frac{1}{(\infty)} + \frac{1}{(498)} = \frac{1}{f} \Rightarrow f = 498$$

The focal length is equal to the far point less the distance between the glasses and the eye. (For a contact lens the distance between the lens and the eye is negligible.) The focal length is 500 centimeters. To calculate the refractive power, we take the reciprocal of this focal length when measured in meters: $1/(5.0) = 0.2$ diopters.

Wave Nature Of Light

We now move on to physical optics. This subject is very involved; we will only touch the highlights.

Young's double slit experiment establishes the wave nature of light. Light passes through a single slit then through a double slit.[3] The pattern on a screen shows a characteristic pattern of bright and dark bands. The distance between the bands is governed by the following equation:

$$\sin\theta = m(\lambda/d) \tag{5.14}$$

where m is the bright fringe index ($m = 0, 1, 2, \ldots$), and d is the distance between the slits.

---

[3] The single slit guarantees that the light through the double slit is in phase.

Wave Nature Of Light

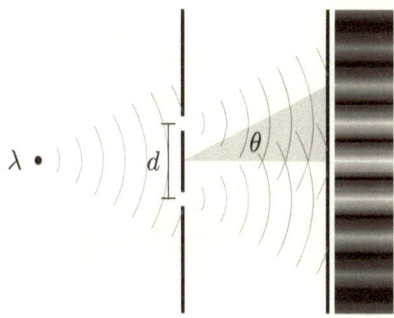

Figure 5.4: Young's double slit experiment

Wave Interference – Light

Consider Young's double slit experiment using a slit distance of 0.23 millimeters. The distance between the double slit and the screen is 4.0 meters. One measures the distance between the center and the first bright fringe to be 0.010 meters. What's the wavelength of the light?

Answer: 570 nanometers

We have everything we need from our problem to use this equation except θ. But we are only a little bit of trig away from that:

$$\tan\theta = \frac{\text{opp}}{\text{adj}} = \frac{(0.010)}{(4.0)} \Rightarrow \theta = 0.143°$$

Therefore,

$$\sin\theta = m(\lambda/d)$$
$$\Rightarrow \sin(0.143°) = (1)(\lambda/(2.3 \times 10^{-4}))$$
$$\Rightarrow \lambda = 5.7404 \times 10^{-7}$$

This wavelength of light is in the yellow-green portion of the visible spectrum.

If we remove one of the slits from the double slit experiment, we have diffraction. This is the tendency of waves to bend around obstacles. The pattern and formula for diffraction bands are similar to Equation (5.14):

$$\sin\theta = m(\lambda/W) \tag{5.15}$$

Where W represents the width of the slit.

The first dark band around the central intensity peak in the diffraction pattern forms a natural "border" for the image of the object. Two objects become indistinguishable when these borders overlap. For a circular hole, the angle corresponding to this first fringe is given by

$$\sin\theta = (1.22)(\lambda/D) \tag{5.16}$$

where D is the diameter of the hole. The extra factor of 1.22 is because of the circular geometry of the slit. This means that the minimum angular separation is given by

$$\theta \approx 1.22\lambda/D$$

It is common to drop the sine since the angle is usually very small (this *only* works if the angle is measured in radians). But with modern calculators it is simple to invert the sine function. Then you don't have to worry so much about degrees versus radians, etc. Just be consistent.[4]

---

[4] This is somewhat blasphemous. Your professor might require you to use radians for such a problem. If so, use them!

Wave Nature Of Light

> **Resolving Power**
>
> The Moon orbits the Earth at an average distance of 3.84×10^8 meters. Considering visible light to have an average wavelength of 600 nanometers, what is the maximum distance one can travel in space before the two objects are indistinguishable to the naked eye?
>
> **Answer:** 1.6×10^{12} meters

In our problem, obviously $\lambda = 6.0 \times 10^{-7}$ meters, but what is D? This is the slit distance that is causing the diffraction. Where is the slit in this problem? Your eye. And the screen upon which the diffraction occurs? Your retina. Assuming your pupil has a diameter of 3.0 millimeters, we have:

$$\sin\theta = (1.22)(6.0 \times 10^{-7})/(3.0 \times 10^{-3}) = 0.0140°$$

To figure out the distance that causes this angle, we need some trig. Since tangent is opposite over adjacent.

$$\tan(0.0140°) = (3.84 \times 10^8)/x \Rightarrow x = 1.5715 \times 10^{12}$$

This is about 10.5 AU. (An "astronomical unit" is the average distance between the earth and the sun).

Interference can also occur in the reflection and refraction of a thin film. The refracted beam enters the film and is reflected off of the back surface. When it returns to the front surface, it interferes with the original reflected beam. This causes the spectacular pattern of soap bubbles, for instance.

Thin Film

A glass plate sits at the slightest angle over a reflective surface. The angle is created by placing a piece of paper under one edge of the 0.10 meters long plate. Coherent light of wavelength 700 nanometers shines directly down. If the distance between the interference bands is 1.0 millimeters, what is the thickness of the paper?

Answer: 3.5×10^{-5} meters

The interference is caused between the ray of light reflected off the back side of the glass plate and the ray of light reflected off the reflective surface.[5]

Since the waves are in phase going down, the interference is caused by the incremental path-length in the gap. When the second ray gets back up to the glass plate it has traveled a distance of $2x$ (where x is the gap distance) and therefore is out of phase by $2x/\lambda$ wavelengths. In this case λ is 700 nanometers because the gap is in air.[6]

Now, when this phase difference is 0°, the interference is constructive. When the phase difference is 360°, the interference is constructive again. A phase difference of 360° corresponds to one wavelength. So, we expect the interference pattern to repeat every time the phase difference is equal to one wavelength. But we have just determined that the phase difference is $2x/\lambda$. Therefore, the gap distance required to repeat the interference pattern is:

$$2x/\lambda = m \Rightarrow x = 3.5 \times 10^{-7}$$

where we have used $m = 1$ to isolate the first interference band.

---

[5] Assume the ray of light reflected off the front of the glass plate is very far away relative to the gap between these two rays.

[6] If the gap were filled with some fluid or something we would need to use the wavelength of the light in the fluid.

Wave Nature Of Light

In the problem there are 100 interference bands (0.10 meters / 1.0 millimeter). So the height at the paper is $100x = 3.5 \times 10^{-5}$.

There is an obscure fact in the middle of this problem. It doesn't affect the answer, but for the sake of completeness I should mention it. When light reflects off of a surface with a higher index of refraction (like from air to glass), the phase of the wave is shifted 180°, or half a wavelength. This can be justified by considering light as a wave and recognizing the distinction between "hard" and "soft" reflection. But it might be simpler to just take it as an empirical fact.

This means the formula above should read

$$2x/\lambda + 1/2 = m$$

Remember that we are given "the distance between the interference bands". This means we are looking for the differences in this formula, i.e.,

$$2\Delta x/\lambda = \Delta m$$

Use $\Delta m = 1$ and we get the same result as before.

In a double slit experiment, both diffraction and interference are working. Diffraction causes the waves to spread out and interference creates the particular pattern on the screen. This interference can be determined by looking at the differences in path-lengths from the two slits. If the difference is an integer number of wavelengths, the interference is constructive. If the difference is a half-integer number of wavelengths, the interference is destructive. Differences in between have interference in between.

A diffraction grating is the double slit experiment on steroids. A typical diffraction grating may have thousands of slits per centimeter. One can analyze the interference pattern the same way. In the end one finds that the constructive interference peaks are separated by

$$\sin \theta = m\lambda/d \qquad (5.17)$$

where d is the slit distance. This is the same formula as that for a double slit. One finds however that the peaks are much sharper and thinner than with the double slit.

Diffraction Grating

X-rays of wavelength equal to 5.0×10^{-11} meters shine on a particular crystal. If the distance between the first point of light from the central beam is $30°$, what is the inter-atomic distance between the atoms in the crystal?

Answer: 1.0×10^{-10} meters

We are going to use our diffraction grating Equation (5.17). In this case $\theta = 30°$ and $\lambda = 5.0 \times 10^{-11}$ m. Therefore

$$\sin(30°) = (1)(5.0 \times 10^{-11})/d \Rightarrow d = 1.000 \times 10^{-10}$$

This is one of the ways that atomic structures are analyzed.

CHAPTER SIX

ELECTRO-MAGNETISM

It is not hyperbole to say that modern civilization is founded on the harnessing of electric power. As a consequence, it is worthy of some study in the physics curriculum. Electricity is also fundamental because it is the source behind all of the other macroscopic forces we see around us (except gravity). Elastic forces, contact forces, support forces, capillary action, surface tension. All of these forces are fundamentally electric in origin because the atom is an object that carries electric charge.

Until now we have been discussing mechanics and various mechanical models of matter. The subject of electromagnetism shifts the discussion. In history, as the study of electricity and magnetism matured, different mechanical models were proposed to explain the phenomena. Strain in an electric ether, vortices in a fluid. Even Maxwell tried to justify his equations using gears and cogwheels.

This approach was doomed to failure. The key to unlocking a complete understanding of electromagnetic phenomena was a new idea: the idea of a field. Its importance was not so much in defining a new model, but in freeing the mind from the classical ones. This new concept focused attention on discovering which laws or equations captured the way the field changed. This was important because it paved the way to seeing beyond Newtonian mechanics into relativity.

So we start with a study fields and how to use them. Unfortunately, we will only be able to scratch the surface. The mathematics becomes quite involved. But we can introduce the fundamentals.

Field From Multiple Charges

Coulomb's law states that the force between two charges is[1]

$$F = k\frac{Qq}{r^2} \qquad (6.1)$$

where q is the electric charge measured in the SI unit of the *coulomb* and the value of k is 8.99×10^9. Since the force is proportional to the electric charge, it is possible to define the electric field in the following way:

$$E = \frac{F}{q} = k\frac{Q}{r^2} \qquad (6.2)$$

In other words, the electric field is the force per unit test charge.

---

[1] This is the same form as Newton's Law of Gravity. The conceptual history of the electric force and gravity is quite intertwined. First, Coulomb modeled his equations after Newton. Later, Einstein modeled his equations (general relativity) after Maxwell.

Field From Multiple Charges

> **Superposition – Electric Field**
>
> Three charges sit on the corners of a rectangle 3.0 meters by 4.0 meters. The charges in the top-right corner and bottom-left corner are +1 coulomb. The charge in the bottom-right corner is −2 coulombs. What is the electric field at the top-left corner?
>
> **Answer:** 1.6×10^9 newtons at $185°$

Let us label the charge at the bottom-left corner #1, the charge at the top-right corner #2, and the charge at the bottom-right corner #3. The distances from the top left corner are 3, 4, and 5 meters respectively.

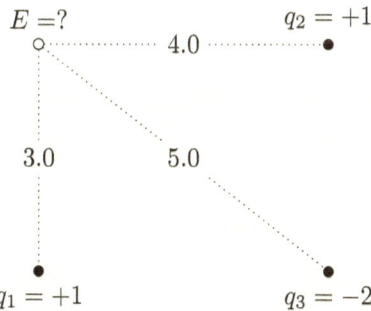

Figure 6.1: Superposition of the electric field from three charges

According to Equation (6.2), we have

$$E_1 = k(+1)/(3)^2 = 9.9889 \times 10^8$$
$$E_2 = k(+1)/(4)^2 = 5.6188 \times 10^8$$
$$E_3 = k(-2)/(5)^2 = -7.1920 \times 10^8$$

The next step is to add together these contributions to the electric field. Remember that the field is nothing more than the electric force per charge. This means that the field is also a vector. The direction of the field is away from the charge because positive charges repel. The field points toward a negative charge, but that is captured in the negative sign.

So what we have above are the magnitudes of these vectors. The angles associated with them are:

$$\theta_1 = 180°$$
$$\theta_2 = 270°$$
$$\theta_3 = \tan^{-1}(-3/4) \approx -36.9°$$

In order to add these vectors together, we need to calculate their components.

$$\begin{aligned} E_{1,x} &= E_1 \cos\theta_1 \\ &= (9.9889 \times 10^8)\cos(180°) = -9.9889 \times 10^8 \\ E_{2,x} &= E_2 \cos\theta_2 \\ &= (5.6188 \times 10^8)\cos(270°) = 0 \\ E_{3,x} &= E_3 \cos\theta_3 \\ &= (-7.1920 \times 10^8)\cos(-36.9°) = -5.7513 \times 10^8 \end{aligned}$$

which adds up to $E_x = -1.5741 \times 10^9$. As for the y component:

$$\begin{aligned} E_{1,y} &= E_1 \cos\theta_1 \\ &= (9.9889 \times 10^8)\sin(180°) = 0 \\ E_{2,y} &= E_2 \cos\theta_2 \\ &= (5.6188 \times 10^8)\sin(270°) = -5.6188 \times 10^8 \\ E_{3,y} &= E_3 \cos\theta_3 \\ &= (-7.1920 \times 10^8)\sin(-36.9°) = 4.3152 \times 10^8 \end{aligned}$$

Field From Multiple Charges

which adds up to $E_y = -1.3034 \times 10^8$. The magnitude of the field is therefore

$$\text{mag } \vec{E} = \sqrt{(E_x)^2 + (E_y)^2}$$
$$= \sqrt{(-1.5741 \times 10^9)^2 + (-1.3034 \times 10^8)^2}$$
$$= 1.5795 \times 10^9$$

and the angle is

$$\text{ang } \vec{E} = \tan^{-1} \frac{E_y}{E_x}$$
$$= \tan^{-1} \frac{-1.3034 \times 10^8}{-1.5741 \times 10^9}$$
$$= 185°$$

Remember, since both components are negative the direction must point into the third quadrant. The final answer is 1.5795×10^9 N at $185°$.

This is a huge force. This is because the charges in the problem are huge. It is much more typical to have micro-coulombs of charge.

Now for a problem that is a little beyond the scope of this book, but I think it is important to cover.

> ### Electric Dipole – Far Field
> Show that the electric field far away from a dipole is proportional to $1/r^3$.
>
> **Answer**: See below

A dipole consists of two equal but opposite charges separated by a small distance. Let's denote the magnitude of the charges as q and the distance between them d. Let the direction from the negative to the positive charge by the positive y direction (up).

In addition, let the distance from a point far away from the dipole and the center of the dipole be r. The angle of this distance will be θ. That's enough set up. Now for the solution.

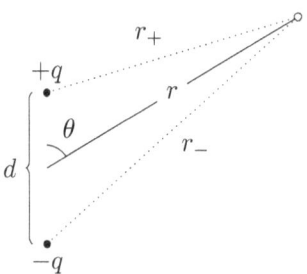

Figure 6.2: Far-field calculation for an electric dipole

Imagine the three lines between the far point and three points on the dipole: the positive charge (on top), the center, and the negative charge (on bottom). As the far point gets farther away, these three lines become more parallel. Since we are looking for the electric field "far away", we will assume these lines are parallel.

Now imagine a line drawn through the center of the dipole perpendicular to these parallel lines. At the top there is a gap while at the bottom the perpendicular crosses the parallel line. These two distances are the same. Let's call it x. By the geometry of the problem we have

$$x = \tfrac{1}{2} d \sin \theta$$

We have already agreed to call the distance from the center to the far point r. The distance between the positive charge and the far point is $r - x$. The distance between the negative charge and the far point is $r + x$. Therefore the two contributions to

Field From Multiple Charges

the electric field at the far point are:

$$E_{\text{pos}} = \frac{kq}{(r-x)^2}$$

$$E_{\text{neg}} = -\frac{kq}{(r+x)^2}$$

Since we have assumed these lines are parallel, the direction these field contributions point are the same. So the total field at the far point is

$$E = \frac{kq}{(r-x)^2} - \frac{kq}{(r+x)^2}$$

Which we can rewrite as

$$E = \frac{kq}{r^2}\left[\frac{1}{(1-x/r)^2} - \frac{1}{(1+x/r)^2}\right]$$

I have chosen to write it this way because we are about to pull out something from our bag of tricks. It is

$$(1+\delta)^n = 1 + n\delta \text{ when } \delta \ll 1$$

This formula is so useful that you will see it used again and again in various places. In our case, we have

$$x/r \ll 1 \Rightarrow (1 \pm x/r)^{-2} = 1 \mp 2x/r$$

This allows us to write the electric field from above as

$$E = \frac{kq}{r^2}\left[\left(1 + \frac{2x}{r}\right) - \left(1 - \frac{2x}{r}\right)\right]$$

This simplifies to

$$E = \frac{(kq)(4x)}{r^3}$$

but since $x = \frac{1}{2}d\sin\theta$, we have

$$E = \frac{(kq)(2d\sin\theta)}{r^3}$$

The *dipole moment* is defined as

$$p = qd \qquad (6.3)$$

which captures the overall distribution of charge. Thus,

$$E = \frac{2kp}{r^3} \sin\theta$$

Which shows explicitly that the dependence is as $1/r^3$. Notice there is also a strong angular dependence. The far-field perpendicular to the dipole is zero.

The dipole is important because it is the simplest way to model the way atoms interact electromagnetically. The dipole shows that an electrically neutral object (like an atom) can generate an electric field. In a rough way, this is how electromagnetism explains the various macroscopic forces of elasticity, friction, etc.

In addition, the simplest type of antenna is a dipole antenna. The electromagnetic radiation pattern follows this electric field pattern.

Electric Potential

The concept of field is closely related to force. But the electric force is a conservative one, so we can also talk about electric potential energy. The corresponding concept in field language is the *field potential*. It follows that the definition of the electric potential is

$$\Delta V = -E \Delta x \qquad (6.4)$$

This equation only works when E is constant. The negative sign is really a hold-over from the definition of gravitational potential. It guarantees that potential decreases as one falls *into* the field.

Electric Potential

The electric potential from a single charge is

$$V = -kQ/r \qquad (6.5)$$

where the reference potential (i.e., $V = 0$) is infinitely far away. The electric potential is measured in volts.

Superposition – Electric Potential

Three charges sit on the corners of a rectangle 3.0 meters by 4.0 meters. The charges in the top-right corner and bottom-left corner are $+1$ coulomb. The charge in the bottom-right corner is -2 coulombs. What is the electric potential at the top-left corner?

Answer: 1.6×10^9 volts

Since the potential is not a vector all we need do is add the three contributions from each charge:

$$V_1 = -k(+1)/(3) = -2.9967 \times 10^9$$
$$V_2 = -k(+1)/(4) = -2.2475 \times 10^9$$
$$V_3 = -k(-2)/(5) = 3.5960 \times 10^9$$

We simply add:

$$V = 1.6482 \times 10^9$$

Clearly we should calculate answers using electric potential rather than the electric field whenever possible.

Accelerating An Electric Charge

The voltage difference in a standard CRT is about 10,000 volts. If an electron starts at rest, how fast is it traveling when it reaches the other end of the screen?

Answer: 5.9×10^7 m/s

The potential energy of the electron is

$$qV = (1.60 \times 10^{-19})(10,000) = 1.60 \times 10^{-15}$$

When dealing with elementary particles it is common to use a smaller non-SI unit of energy: the *electron-volt*. This is the amount of potential energy an electron has in one volt of electric potential. In other words,

$$1 \text{ eV} = 1.60 \times 10^{-19} \text{ J}$$

For our problem we could say the electron has a potential energy of 10,000 eV. Since we will need to convert this to SI units to solve the problem anyway, we will leave it as it is now.

The particle starts at rest, so the initial energy is all potential. At the end of the electric field, all this energy is kinetic. Thus,

$$1.60 \times 10^{-15} = \tfrac{1}{2}mv^2 = \tfrac{1}{2}(9.11 \times 10^{-31})(v^2)$$
$$\Rightarrow v = 5.9267 \times 10^7$$

This is nearly 20% the speed of light! In this problem we need to take relativistic effects into account. But we will leave that for the next chapter.

The Capacitor

The simplest situation in which to discuss the electric potential is the *capacitor*. A capacitor is two parallel conductive plates that are separated by a small distance. This gap may either be filled with an insulator or by vacuum. Charge trapped on the plates cannot cross the gap. As positive charges accumulate on one plate, negative charges on the other plate are attracted but cannot cross. This sets up a situation in which the electric charges are "captured" in the device. Thus, capacitors store electric charge.

The Capacitor

Of course, the charges set up an electric field between them. If the plate area is large compared to the gap distance, it can be shown by symmetry that the field everywhere inside the capacitor is constant:

$$E = 4\pi k Q/A$$

where Q is the total charge (on the positive plate) and A is the total area of the capacitor plate.

Because the field is constant, the potential is easy to calculate:

$$V = -4\pi k (Q/A) x$$

where x is the distance from the positive plate. So the potential is zero at the positive plate and becomes more negative as one gets closer to the negative plate. The potential energy of a charge q caught in this capacitor is qV.

The ratio of the total charge held by a capacitor and the voltage between the plates is called the *capacitance* of the capacitor (SI unit is the *farad*). For our ideal capacitor we have

$$C = Q/V = \epsilon_0 d/A \qquad (6.6)$$

where we have used $\epsilon_0 = 1/4\pi k$. This formula assumes the capacitor is very large. Another way to say this is that it ignores edge effects. On the edge of the capacitor, the field will not be constant, so the analysis fails. Thus, the capacitance of any real capacitor will vary from this ideal value.

Building a Capacitor

A capacitor is constructed with two conducting plates (area $= 1.0$ cm^2) separated by a distance of 0.10 millimeters with an insulating material. When 10 volts are held across the capacitor, the total charge held is 2.0×10^{-10} coulombs. What is the dielectric constant of this material?

Answer: 2.26

From the definition of capacitance we have

$$C = Q/V = \kappa \epsilon_0 d/A$$

In this case, κ is the dielectric constant of the material. When we plug in the data we get:

$$(2.0 \times 10^{10})/(10) = (\kappa)(8.854 \times 10^{-12})(0.0001)/(0.0001)$$
$$\Rightarrow \kappa = 2.26$$

This is not really a good way to measure the dielectric constant because this formula is dependent on ignoring the edge effects on the capacitor.

Ohm's Law, Power Loss

Suppose we have a charged capacitor. If we connect the two plates with a wire, the charges will flow and cancel one another out. This flow of charge is called *current*:

$$I = \Delta Q/\Delta t \tag{6.7}$$

The SI unit of current is the *amp*. Of course as the capacitor discharges, the amount of current drops. A battery is like a capacitor in that it holds a potential difference, but with enough charge that the drop in current is negligible.[2]

Thus, a battery will drive current around a circuit. In general, there is some relationship between the driving voltage and the resulting current. The relationship is either linear or not. If the relationship is non-linear, the circuit element is called "active". Examples include diodes and transistors. If the relationship is linear, we have a *resistor*. So, by definition,

$$V = IR \tag{6.8}$$

---

[2] Batteries do wear out, of course, it just takes a "long" time.

Ohm's Law, Power Loss

where R is the *resistance* of the resistor (SI unit is the *ohm*). This is known as Ohm's law. All resistors have some voltage range over which Ohm's law is valid. Good ones have a wide range of validity.

> **Current and Charges**
>
> A 9.0 volt battery is placed across a resistor whose value is 1500 ohms. How many electrons run through the battery each second?
>
> **Answer:** 3.8×10^{16} electrons

The number of electrons per second is related to the current. We can use Ohm's Law to calculate this current:

$$V = IR \Rightarrow (9.0) = (I)(1500) \Rightarrow I = 0.0060$$

or 6.0 milliamps. This is a typical value—an amp is pretty large. We are interested in a timeframe of one second, so the total amount of charge that flows is given by Equation (6.7):

$$0.0060 = (\Delta Q)/(1) \Rightarrow \Delta Q = 0.0060$$

Since each electron has a charge of 1.60×10^{-19} coulombs, the total number of electrons involved must be

$$0.0060/1.60 \times 10^{-19} = 3.750 \times 10^{16}$$

Normally, an electric potential difference will accelerate charge. But if a resistor is in the path of this potential difference, the resistor opposes the acceleration. The charge moves, but at a constant speed[3] (i.e., current). Therefore resistance is a loss of energy (into friction and heat). The amount of power lost to heat is

$$P = VI \qquad (6.9)$$

---

[3] This is a kind of like terminal velocity in free-fall.

Electric Power Loss

In the previous problem, how much energy does the battery lose each second?

Answer: 14 kilojoules

This is simple. Multiply the voltage and the current via Equation (6.9):

$$P = (9.0)(1500) = 14000$$

Since power is energy per second, we have

$$E = Pt = (14000)(1) = 14000$$

One last problem. What if we combine a resistor and a capacitor?

RC Circuit

Suppose a fully charged capacitor, with a capacitance of 10 microfarads, carries a voltage of 5.0 volts. If we place a 500 ohm resistor across the plates, how long does it take for the capacitor to discharge 95% of its initial charge?

Answer: 15 milliseconds

This is called an RC circuit. Initially, current will flow:

$$I_0 = V_0/R = (5.0)/(500) = 0.010$$

The amount of charge on the capacitor is

$$Q_0 = CV_0 = (5.0)(10 \times 10^{-6}) = 5.0 \times 10^{-5}$$

Since $I = \Delta Q/\Delta t$, you might think we can find t by

$$\Delta t = \Delta Q/I = (5.0 \times 10^{-5})/(0.010) = 0.005$$

or 5.0 milliseconds. But this is not correct. As the current flows and the capacitor discharges, the voltage across the capacitor drops. This means less current is driven around the circuit and less charge flows.

The formula for the discharge is:

$$Q(t) = Q_0(1 - \exp(-t/RC))$$
$$I(t) = I_0(1 - \exp(-t/RC))$$
(6.10)

According to the formula, the capacitor never fully discharges. We are looking for the time at which the left-over charge is 95% of the original. In symbols:

$$Q(t) = (0.95)Q_0$$

When we substitute this in the above equation we get:

$$(0.95)(Q_0) = (Q_0)(1 - \exp(-t/RC))$$
$$\Rightarrow \exp(-t/RC) = 0.05$$
$$\Rightarrow t = -RC\ln(0.10) = (3.00)(RC)$$

The quantity RC is called the *time constant* for the RC circuit. This shows that the capacitor is 95% discharged after three time constants. In our case, the time constant is

$$RC = (500)(10 \times 10^{-6}) = 0.005$$

Which is the same number we calculated earlier. But now we know that the answer is three times this number:

$$t = (3.0)(0.005) = 0.015$$

The final answer is 15 milliseconds.

Equivalent Resistance

If an electric circuit is connected such that all the current must flow through all the elements, these elements are said to be a *series circuit*. If they are connected such that the same voltage is applied across them, these elements are said to be a *parallel circuit*.

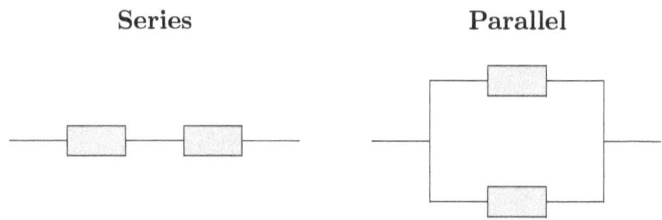

Figure 6.3: Components in series or parallel

Resistors in series add, but in parallel their reciprocals add:

$$\text{Series: } R = R_1 + R_2$$

$$\text{Parallel: } \frac{1}{R} = \frac{1}{R_1} + \frac{1}{R_2}$$

For capacitors, the rules are reversed:

$$\text{Series: } \frac{1}{C} = \frac{1}{C_1} + \frac{1}{C_2}$$

$$\text{Parallel: } C = C_1 + C_2$$

Consider the following.

Equivalent Resistance

A 300 ohm resistor is in parallel with a 600 ohm resistor. In series with this pair is a 200 ohm resistor. A voltage of 6.0 volts is applied. How much current runs through the 600 ohm resistor?

Answer: 5.0 microamps

Equivalent Resistance 167

The current from the battery runs through the 200 ohm resistor and is then split into the pair. The amount that gets split into the 600 ohm resistor is what we are being asked to find. The first step is to find the total current. Since we know the total voltage, we should determine the total resistance.

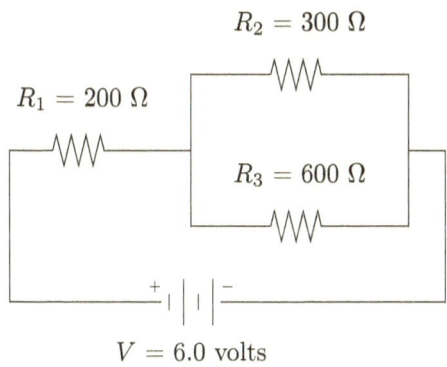

Figure 6.4: Equivalent resistance problem

The equivalent resistance of the pair in parallel is given by:

$$\frac{1}{R} = \frac{1}{300} + \frac{1}{600} \Rightarrow R = 200$$

Now this is in series with the other resistor, so the total resistance of the circuit is:

$$R = (200) + (200) = 400$$

Using Ohm's Law, the total current is

$$I = V/R = (6.0)/(400) = 0.015$$

This current flows through the 200 ohm resistor. By Ohm's Law, the voltage drop across this resistor is

$$V = (0.015)(200) = 3.0$$

This leaves 3.0 volts of voltage to drop across the pair. You can see this by either subtracting from the total voltage, or by applying Ohm's Law to the equivalent resistance of the parallel pair.

Since both parallel resistors have 3.0 volts across them, the total current through the 600 ohm resistor is

$$I = V/R = (3.0)/(600) = 0.0050$$

Unfortunately, not all circuits can be analyzed this way. It is possible to set up a circuit in which the components are neither in series nor parallel. The simplest example is the Wheatstone Bridge. Imagine a square with resistors on all four sides. Connect one of the pairs of opposite corners with a voltage. Place a fifth resistor (or an ammeter) between the other pair of opposite corners. There are no pair of resistors which are in series or parallel.

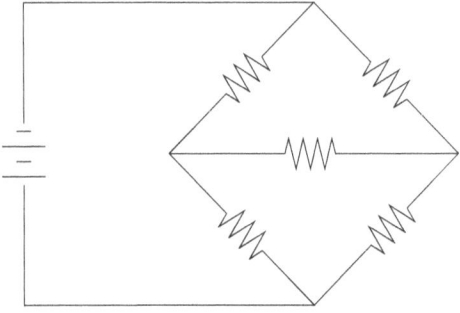

Figure 6.5: Wheatstone Bridge

If one places an ammeter in the center, one can get a very sensitive measurement of the relative resistances on the square. This is the electrical equivalent of a regular mass balance.

The way to analyze a more complicated circuit like this is to use Kirchhoff's laws:

- At every circuit junction, the total current flowing in equals the total current flowing out.

- Around every circuit loop, the total voltage dropped is equal to the total voltage applied.

These are derived from the conservation of charge and the conservation of energy, respectively.

This method requires one to set up an equation for each junction and each loop in the circuit. For example, in the Wheatstone Bridge, there are four junctions and three loops. This is a total of seven equations. Skills in linear algebra are helpful for problems like these.

Magnets, Motors, And Solenoids

In 1785, Coulomb performed experimentation to quantify the magnetic force and discovered that the ends (or poles) of the magnet obey an inverse square law (similar to electric charge):

$$F = \frac{\mu_0}{4\pi} \frac{pP}{r^2} \qquad (6.11)$$

where p represents the strength of the poles and the constant μ_0 represents the *permeability* of vacuum (in SI units: 1.27×10^{-6}).

Now, the poles of a magnet are not actually located at the ends of the magnet. Rather they are somewhat inside (about 1/6 from the end or so). If this were not true, we would never be able to pull magnets apart! This imprecision introduces a practical limitation on Coulomb's Law for magnets.

Superposition – Magnetic Field

Consider a magnet with poles separated by 0.10 meters. What is the magnetic field 0.12 meters away in the plane perpendicular to the line connecting the poles? Assume the pole strengths to be 4.0×10^{-7} units.

Answer: 27.7 tesla

The distance between each pole and the measurement point is 0.13 meters. It just so happens that the geometry involves a 5-12-13 triangle. (You can thank me later.) So, according to Equation (6.11) the magnitude of the magnetic field from each pole is

$$B = \frac{\mu_0}{4\pi} \frac{P}{r^2}$$

This is because the magnetic field is defined through the equation $F = pB$ where p represents the magnitude of the pole experiencing the magnetic force. Just as we did with the definition of the electric field.

In our case, we have

$$B = \frac{\mu_0}{4\pi} \frac{P}{r^2}$$
$$= \frac{4.0 \times 10^{-7}}{(4\pi)(1.257 \times 10^{-6})(0.13)^2}$$
$$= 14.98$$

The angle associated with this field is given by

$$\theta = \tan^{-1}(5/12) = 22.62°$$

We need to calculate the components of the magnetic field. Let's align the y-axis with the two poles of the magnet. Then the components are:

$$B_x = (B)(\cos\theta) = (14.98)(\cos 22.62°) = 13.828$$
$$B_y = (B)(\sin\theta) = (14.98)(\sin 22.62°) = 5.7616$$

Magnets, Motors, And Solenoids

The y-components cancel and the x-components add. So, the final field from both poles point directly away from the dipole in the x-direction with a magnitude of

$$2B_x = (2)(13.828) = 27.656$$

This is what we were asked to solve.

The smallest element of magnetism observed physically is not a magnetic pole but a magnetic *dipole*. If one splits a magnet in two, one gets two magnets not two poles. Each molecule and atom (even elementary particles like the electron) have some sort of magnetic moment. We will see later that this is due to the motion of the electric charges contained within them.

As a consequence, the "magnetic pole" approach to magnetism is used infrequently. Rather the magnetic moment of the dipole is considered the primary quantity. These concepts are connected through the definition of the magnetic moment:

$$m = ps \qquad (6.12)$$

Compare this with the equation for the electric dipole moment in Equation (6.3). Notice that this is a vector equation. The direction of the magnetic moment points from the south to the north pole of the magnet.

Suppose we have a bar magnet and the north pole of another magnet is brought near. The north pole of the first magnet will experience a repulsive force while the south pole will experience an attractive force. This pair of forces will tend to twist the magnet. In other words, the magnet experiences a net torque.[4]

The formula for this torque is given by

$$\tau = mB \sin\theta \qquad (6.13)$$

---

[4] This torque guarantees that two magnets will always be attracted to one another, even if they have to twist around first.

where θ is the angle between the magnetic moment and the magnetic field.

> **Magnetic Deflection**
>
> A compass with a magnetic moment of 1.0 unit is oriented at NNE relative to the Earth's 50 microtesla magnetic field. Calculate the torque applied to this compass.
>
> **Answer:** 1.9×10^{-5} newton-meters

The angle here is the only real tricky part. An orientation of NE is 45°, so the orientation of NNE is 22.5°. Therefore the total torque, given by Equation (6.13), is

$$\tau = (1.0)(5.0 \times 10^{-5})(\sin 22.5°) = 1.9134 \times 10^{-5}$$

An electric charge at rest in a magnetic field feels no force. But put the charge in motion and—*voila*—there is a force. The formula is:

$$F = qvB \sin \theta \qquad (6.14)$$

This is the magnitude of the force where θ is the angle between the magnetic field and the velocity vectors.

The direction of this force is perpendicular to the plane that holds the magnetic field and the velocity vectors. This leaves one of two options for the direction. By convention, we must use the *right hand rule* to choose between them. Point your index finger in the direction of the charge motion and curl your other fingers to point in the direction of the field then your thumb points in the direction of the force.

Magnets, Motors, And Solenoids

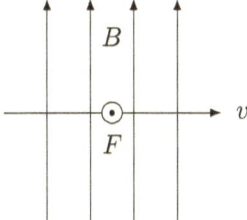

Figure 6.6: The magnetic force on this moving electric charge points out of the paper

Lorentz Forces

An electron is travelling east at a speed of 100 m/s. It encounters a magnetic field pointing north. What magnitude of magnetic field is required to overcome the acceleration due to gravity, effectively levitating the electron?

Answer: 5.57×10^{-13} tesla

The force experienced due to gravity is simply the weight:

$$W = mg = (9.109 \times 10^{-31})(9.8) = 8.927 \times 10^{-30}$$

The magnetic force needs to counter-balance this force. The magnitude of the magnetic force is given by Equation (6.14). In this case:

$$(8.927 \times 10^{-30}) = (1.602 \times 10^{-19})(100)(B)$$
$$\Rightarrow B = 5.57 \times 10^{-13}$$

This is a miniscule amount of field. This shows how much more powerful the electromagnetic force is than gravity.

Since the force of the magnetic field is perpendicular to the motion of the particle, the field does no work on the particle. Its speed does not change, only its direction. Thus, a charged particle moving in a magnetic field will move in a circle perpendicular to the field.[5]

The centripetal force required to maintain circular motion is $F = mv^2/r$. When we assume this centripetal force is provided by the magnetic field, we have,

$$mv^2/r = qvB \Rightarrow r = mv/qB$$

This is the principle behind the *mass spectrometer*. A sample to be analyzed is pushed into a magnetic field and the radius of the curvature gives a measurement of the mass of the particle (assuming one knows the charge). This can be used to isolate different isotopes or simply determine the composition of an unknown substance.

Mass Spectrometer

A certain mass spectrometer is designed with a radius of 0.50 meters using a magnetic field of 2.0 tesla. What electric voltage is required to accelerate a doubly ionized ozone molecule to travel through the spectrometer?

Answer: 4.0 million volts

We need to determine the velocity of the ozone molecule when it enters into the spectrometer. Since the ozone molecule is three oxygen atoms, the mass is $m = (3)(15.9994) = 47.998$ u.

---

[5] If the particle has some component of velocity parallel with the magnetic field, the force does not affect this component. So, in general, a charged particle moving in a magnetic field will move in a spiral. This happens in the *aurora borealis*. Charged ions moving toward the earth are deflected by its magnetic field. The particles tend to spiral down the magnetic field lines to the north or south magnetic poles. Since the field actually constricts toward the pole, the spirals become tighter until they collide with molecules in the atmosphere, causing the particles to radiate. Hence the "Northern Lights."

Magnets, Motors, And Solenoids

The conversion to SI units is:

$$(47.998 \text{ u}) \times \left(\frac{1.6605 \times 10^{-27} \text{ kg}}{1 \text{ u}}\right)$$
$$= 7.9701 \times 10^{-26} \text{ kg}$$

Using the formula above yields

$$r = \frac{mv}{qB}$$
$$\Rightarrow (0.50) = \frac{(7.9701 \times 10^{-26})(v)}{(2)(1.602 \times 10^{-19})(2.0)}$$
$$\Rightarrow v = 4.020 \times 10^6$$

This is a non-relativistic speed, so we can use the classical kinetic energy formula:

$$KE = \tfrac{1}{2}mv^2$$
$$= \tfrac{1}{2}(7.9701 \times 10^{-26})(4.020 \times 10^6)^2$$
$$= 6.4401 \times 10^{-13}$$

This incoming kinetic energy comes from the acceleration due to the electric potential energy of the charge:

$$PE = qV$$
$$\Rightarrow (6.4401 \times 10^{-13}) = (1.602 \times 10^{-19})(V)$$
$$\Rightarrow V = 4.020 \times 10^6$$

This circular motion is also used in bubble chambers. A *bubble chamber* is a supersaturated gas into which particles are injected. The particle leaves behind a trail of droplets which make it easy to reconstruct the path of the particle. If one puts this apparatus in a magnetic field, charged particles will be deflected. This deflection gives information about the mass-to-charge ratio m/q.

Since a current is nothing more than charge in motion, we should expect a wire with current running through it to be deflected in a magnetic field also. And it is. It is not difficult to show that

$$F = ILB\sin\theta \qquad (6.15)$$

where I is the current, L is the length of the wire in the magnetic field, B is the magnetic field, and θ is the angle between the direction of the current and the field. Similarly, a coil of wire will experience a torque given by

$$\tau = IAB\sin\phi \qquad (6.16)$$

where A is the area of the coil and ϕ is the angle between the direction normal to the surface of the coil and the direction of the field. Notice that if we compare this with Equation (6.13), the magnetic moment of a coil is given simply by

$$m = IA \qquad (6.17)$$

If there is more than one coil, multiply by the number of coils.

This torque can be used to build a simple DC motor. We begin by placing a coil inside the magnetic field of a permanent magnet. If we apply a current, the coil will twist. The connection to the battery must be accomplished with some graphite contacts called "brushes" to accommodate the twist. If we make the brushes such that a half turn will switch the current, the torque due to the magnetic field will flip direction—which continues to push the coil in the same direction of turning. Once the coil makes a full rotation, the process repeats. We have built a motor from a battery, wire, and some graphite brushes!

Magnets, Motors, And Solenoids

> **DC Motor**
>
> How much current is required to accelerate a DC motor up to 50 revolutions per second in one second? The motor uses 50 tesla magnets, has a moment of inertia equal to 0.10 kg-m^2, and a radius of 2.0 centimeters. Assume there are 200 turns in the coil of wire.
>
> **Answer:** 2.5 amps

The required acceleration is 50 rev/s^2. In radians this is

$$\alpha = \left(\frac{50 \text{ rev}}{\text{s}^2}\right) \times \left(\frac{2\pi \text{ rad}}{1 \text{ rev}}\right) = 314.16$$

The torque required to produce this angular acceleration is

$$\tau = I\alpha = (0.10)(314.16) = 31.416$$

The current through the electromagnet creates a magnetic moment that twists the motor through the 50 tesla magnetic field. This twisting torque is given by Equation (6.13):

$$\tau = mB \Rightarrow (31.416) = (m)(50) \Rightarrow m = 0.62832$$

Since the magnetic moment of the current is given by Equation (6.17), we can extract the required current after we have calculated the relevant area. Thus,

$$A = \pi r^2 = \pi (0.020)^2 = 1.2566 \times 10^{-3}$$

Finally, the required current must be

$$m = NIA$$
$$\Rightarrow (0.62832) = (200)(I)(1.2566 \times 10^{-3})$$
$$\Rightarrow I = 2.50$$

If a charge can be affected by a magnetic field, it stands to reason that a charge might affect the magnetic field. Oersted is credited with this discovery. He saw that a compass needle will be deflected by a nearby current. This result deepens the connection between the magnetic and electric force.

Ampere summarized these phenomena in the following mathematical rule.

$$\sum_{\text{loop}} B_t \Delta \ell = \mu_0 I \qquad (6.18)$$

Since current causes a magnetic field to swirl around it, this formula relates the size of the current to the total amount of magnetic field circulating around it. The sum in the equation represents any closed loop that encircles the current and B_t represents the component of the magnetic field that is tangent to the curve over which the sum occurs.

Two special cases are simpler to understand. First, an infinitely long, straight wire. The magnetic force curls around the wire. The direction of the curl is determined by another right hand rule: if you point your right thumb in the direction of the current, then your fingers will naturally curl in the same direction as the magnetic field. The magnitude of the magnetic field is given by

$$B = \frac{\mu_0 I}{2\pi r} \qquad (6.19)$$

where r is the distance from the wire. This curling action should conjure up images of circulating fields flowing around the wire. Since we already know that a moving electric charge is affected by a magnetic field, this implies that two wires with flowing current will be attracted to one another. They are indeed. In fact, they have a tendency to twist around and braid themselves together like DNA.

The second special case is an infinitely long coil. The magnetic field still "curls" around the coil. But because of the geometry, the field lines in the center of the coil extend to infinity. In other words, we have a field pattern that looks quite similar to

Magnets, Motors, And Solenoids

that of a bar magnet. This construction is called a *solenoid*. The magnetic field outside the solenoid looks just like a regular bar magnet. The magnetic field inside the solenoid is constant. The magnitude of the magnetic field in the center is given by

$$B = \mu_0 n I \tag{6.20}$$

where n is the number of turns per meter of wire, and I is the current running through the wire. This is your good old fashion electromagnet.

Electromagnet

Suppose a certain car can be lifted off of the ground by a 40 tesla magnetic field. An electromagnet with 50,000 turns and a radius of 5.0 meters is used. The resistivity of the wire is 1.72×10^{-8} Ω-m with a cross-sectional area of 2.0×10^{-6} m^2. What voltage is required to lift the car?

Answer: 8.6 million volts

We will use the equation for a solenoid, Equation (6.20), to determine the current required:

$$(40) = (1.257 \times 10^{-6})(50000)(I) \Rightarrow I = 636.62$$

In order to calculate the voltage, we need to know the resistance of the wire. The resistance is related to the resistivity by

$$R = \rho l / A \tag{6.21}$$

The resistance is larger for a longer wire and smaller for a larger wire. In our case, we are given the area and need to calculate the length. Since the wire coils around a circle of radius 5.0 meters, the total length of the wire is

$$l = (50000)(2\pi)(5.0) = 1.5701 \times 10^6$$

So the total resistance is

$$R = (1.72 \times 10^{-8})(1.5701 \times 10^6)/(2.0 \times 10^{-6}) = 13508$$

Finally, by Ohm's Law, the required voltage is

$$V = IR = (636.62)(13508) = 8.600 \times 10^6$$

Electric Generators

In the last section we talked about using electric current to create a magnetic field. Is it possible to use magnetic fields to generate electric current? The answer is yes. Faraday did most of the ground breaking work on this subject. In fact, Faraday is the father of the "field" concept which was born out of his struggle to understand these very phenomena.

The first thing to note is that a magnetic field alone cannot force current around a wire since magnetic fields only affect charges if they are already in motion. Faraday discovered that *changes* in the magnetic field will do the trick. This is called *induced current* and the implied voltage driving this current is called *electromotive force* or EMF.[6]

There are two ways to change the amount of magnetic field affecting a wire. One is to change the magnetic field. Another is to change the geometry of the circuit.

Imagine a circuit immersed in a magnetic field with a sliding conductor that closes the circuit. As the conductor slides up and down the wires, it encloses more and less of the magnetic field. Faraday discovered that this also induces EMF. This idea of a field "enclosed" by a curve is called *field flux*.[7]

---

[6] Be careful. Electromotive force is not a *force* at all. It is electric potential. The term derives from the fact that the electric potential is what drives or "forces" current around a circuit. We now know it is better to talk about it as an electric potential or voltage, but the term is too well entrenched. Sorry.

[7] One particularly good way of thinking about flux is that it is propor-

Electric Generators

The magnetic flux is defined as

$$\Phi = BA\sin\phi \tag{6.22}$$

where B is the magnetic field, A is the area of the enclosure, and ϕ is the angle between normal to the enclosure surface and the direction of the magnetic field. With the use of magnetic flux, we can now state Faraday's Law:

$$EMF = -N\frac{\Delta\Phi}{\Delta t} \tag{6.23}$$

where N is the number of loops and Φ is the magnetic flux through the surface.

Magnetic Flux

There is a uniform horizontal magnetic field of 10 tesla above a table. A square plate (0.20 m by 0.20 m) sits on a table at in incline of 20°. What is the magnetic flux through the plate?

Answer: 0.14 T-m^2

By definition, the magnetic flux is

$$\Phi = (10)(0.2)(0.2)(\sin 20°) = 0.1368$$

The units of magnetic flux are the weber[8] which is simply defined as a tesla times the meter squared.

It follows from Faraday's law that one can induce EMF by either moving a magnetic field near a stationary coil or by

---

tional to the number of field lines that pass through a surface. The larger the number, the greater the flux.

[8] The symbol for the weber is "W". This is also the symbol for a watt of power. Not to mention work and weight. Sigh. Because of this ambiguity, I am going to avoid using the weber and quote the magnetic flux in units of tesla-meter-squared.

moving a coil near a stationary magnetic field. Most electric generators are built using the second method.

> **Faraday's Law – Constant EMF**
>
> A simple electronic circuit is designed with a rectangular piece of wire. The dimensions of the rectangle are 0.20 meters by 0.40 meters. Suppose one of the 0.20 meter wires is moveable and is pushed to the center of the rectangle in 0.10 seconds without losing contact. The end result is a wire surrounding a square 0.20 meters by 0.20 meters. If the wire is situated in a 0.50 tesla magnetic field perpendicular to the area, what is the EMF generated by the motion of the wire?
>
> **Answer**: 0.20 volts

The initial magnetic flux surrounded by the wire is

$$\Phi = (0.50)(0.20)(0.40)(\sin 90°) = 0.040$$

The final magnetic flux surrounded by the wire is

$$\Phi = (0.50)(0.20)(0.20)(\sin 90°) = 0.020$$

So, the change in flux is $(0.040) - (0.020) = 0.020$ T-m^2. According to Faraday's Law,

$$EMF = (1)\frac{0.020}{0.10} = 0.20$$

> **Faraday's Law – Calculate Magnetic Field**
>
> A square loop of wire 0.040 meter on each side sits perpendicular to a magnetic field. The wire is allowed to spin with an angular speed of 40 radians per second so that it is parallel to the magnetic field and an average EMF of 0.010 V is measured. What is the magnitude of the magnetic field?
>
> **Answer**: 0.25 tesla

Electric Generators

In the final state ("parallel to the magnetic field"), the magnetic flux is zero because $\sin 0° = 0$. The magnetic flux in the initial state is

$$\Phi = (B)(0.040)(0.040)(\sin 90°) = (0.0016)(B)$$

According to Faraday's Law, the change in flux is related to the EMF via

$$EMF = N\frac{\Delta \Phi}{\Delta t} = (1)\frac{-(0.0016)(B)}{\Delta t}$$

We need to know Δt in order to finish the problem. The wire rotates one-quarter of a revolution, which is $\pi/2$ radians. Since it is rotating at 40 radians per second, the duration involved is

$$\omega = \frac{\Delta \theta}{\Delta t}$$
$$\Rightarrow (40) = \frac{\pi/2}{\Delta t}$$
$$\Rightarrow \Delta t = 0.03927$$

Therefore,

$$(0.010) = (1)\frac{-(0.0016)(B)}{0.03927} \Rightarrow B = -0.2454$$

The sign indicates that the magnetic field is pointing down (assuming the current is counter-clockwise), but we are merely asked for the magnitude of the field.

Faraday's law determines the magnitude of the electric potential generated by a changing magnetic flux. But it doesn't tell us the *direction* of the potential. This is similar to the ambiguity resolved by the right hand rule in the Lorentz force law, Equation (6.14). Lenz's law is the rule we need:

> Any induced current will have a direction such that it will induce a second magnetic field to oppose the change in magnetic flux that created the initial current.

If the flux is increasing, the second magnetic field will point in a direction opposite to the first magnetic field. The secondary flux will *subtract from* the increase. If the flux is decreasing, the second magnetic field will point in the same direction as the first magnetic field. The secondary field will *add to* the first field. The secondary induced field is trying to bring the first field back to its original state—it is "resisting" the change.

> **Lenz's Law**
>
> Suppose a conducting ring has its axis vertically oriented. A uniform magnetic field pointing up is turned on. Looking down, will current flow clockwise or counter-clockwise?
>
> **Answer**: Counter-clockwise

The change in flux is negative, so the EMF must attempt to add flux. This will happen if the current moves counter-clockwise. According to the right-hand rule, a current pointing east will have a magnetic field that curls underneath toward the north, up and over toward the south. Thus, in the center of this ring, the bottom portion of the coil will produce a magnetic field in the center that points up. The top portion will also produce a magnetic field pointing up. This is what Lenz's law requires.

This completes all of the basic laws of electromagnetism.[9] A few final notes are in order on some basic machines that use these principles.

An electric generator converts mechanical rotary motion into EMF. Almost anything that produces electric energy has some form of electric generator in it, whether it is a dam or a windmill or the generator in your car's engine.

---

[9] Not quite. We still need Maxwell's displacement current to complete Maxwell's laws. But this is a mathematical detail we can overlook in an introductory text.

The rotating coil of wire is placed in a steady permanent magnetic field. Because of the flux change, Faraday's law predicts that current will flow. Notice that with uniform rotary motion the area presented to the magnetic field varies in a sinusoid pattern. The EMF will therefore also vary in a sinusoid pattern. This is how AC voltage is generated. Which is precisely why AC voltage is so popular. DC voltage is actually created by manipulating AC voltage (except when a battery is used).

The induced EMF for a rotating coil is:

$$EMF = NAB\omega \sin \omega t \qquad (6.24)$$

where ω is the angular frequency of the rotation.

Transformers And Inductors

Remember that voltage through a solenoid creates a magnetic field. Suppose we attach an AC voltage to a solenoid. This should create a *changing* magnetic field that can induce EMF in a second solenoid via Faraday's Law. The particulars are dependent upon the geometry and materials used. This effect is called *mutual inductance*.

The main application of mutual inductance is in a *transformer*. The transformer consists of an iron core around which two coils are wound. The primary coil is connected to an AC generator. The iron core constrains most of the magnetic field (and therefore the flux) into the iron core itself. Thus, when the flux in the primary coil changes, the flux in the secondary coil changes by the same amount. Assuming the same material is used for the coils and the same radius, the difference between the two coils is merely in the *number* of coils. The voltage induced in the secondary coil is related to the number of coils by:

$$V_2 = \frac{N_2}{N_1} V_1 \qquad (6.25)$$

If the number of coils on the secondary side is greater, the transformer is said to "step-up". If the number of coils on the secondary side is less, the transformer is said to "step-down".

However, power (or energy) must be conserved. Since $P = VI$, a transformer that steps-up voltage also steps-down current, and *vice versa*.

> **Transformers**
>
> A power plant generates a voltage of 1200 volts. Suppose a 10:1 step-up transformer is used to transmit the power across a wire with resistance of 2000 ohms. What is the step-down transformer required to produce 120 volts of electricity at the output and how much power is lost due to the resistance in the wire?
>
> **Answer**: (a) 1:100 (b) 72 kilowatts

The step-up transformer creates a transmission voltage of 12,000 volts. So the step-down transformer must be 1:100. The total power lost is given by Equation (6.9). The current in the wire is given by Ohm's Law:

$$V = IR \Rightarrow (12000) = (I)(2000) \Rightarrow I = 6.0$$

So the power loss is

$$P = IV = (6.0)(12000) = 72000$$

It may have already occurred to you that you don't need two coils to have this inductive effect. One coil is sufficient. When AC current is driven through a solenoid, the changing magnetic field produces a back EMF (similar to the electric motor example earlier) in accord with Lenz's law. The amount of back EMF is dependent upon the geometry and materials used. This effect is called *self-inductance* and when viewed from this perspective, a solenoid is called an *inductor*. Just as the capacitor stores energy in the electric field, an inductor stores energy in the magnetic field.

AC Circuits

As mentioned previously, electric generators produce AC voltage according to Equation (6.24). We can re-write this formula in terms of the peak voltage V_0 as

$$V(t) = V_0 \sin(2\pi f t) \qquad (6.26)$$

where f is the frequency of the change in voltage. The current also obeys a sinusoidal pattern as well, but the relationship between the two takes some explaining. For a circuit made of resistors, the current still follows Ohm's Law. Therefore the current is related to the voltage as:

$$I(t) = V(t)/R = (V_0/R)\sin(2\pi f t) = I_0 \sin(2\pi f t) \qquad (6.27)$$

Notice that the power in the circuit is given by

$$P(t) = V(t)I(t) = (V_0 I_0)\sin^2(2\pi f t)$$

The average value of $\sin^2 \theta$ is 0.5, so the average power required to supply the circuit using AC current is:

$$\langle P \rangle = \tfrac{1}{2}(V_0 I_0) \qquad (6.28)$$

How will a capacitor react to AC current? As the voltage across the capacitor increases, the capacitor begins to charge. The charge on the capacitor follows the voltage. However, the current is the *change* in the charge. When the voltage is peaking, the current flow is falling to zero since it is not necessary to supply any more charge to the capacitor. Indeed, the current is said to "lead" the voltage.

An inductor reacts differently because of Faraday's law. For the inductor, the changing voltage drives the flux and therefore

Component	Reactance	Phase Shift
Resistor	R	$0°$
Capacitor	$1/2\pi f C$	$90°$
Inductor	$2\pi f L$	$-90°$

Table 6.1: AC Reactance Of Electrical Components

the current across the inductor so the current "lags" behind the voltage. For both electrical components, the voltage and current are 90° out of phase. As a consequence, there is no power loss across these components. The capacitor stores and releases energy via the electric field, while the inductor stores and releases energy via the magnetic field.

You may have noticed the pattern now. We have three components: resistors, capacitors, and inductors. Each determines a relationship between current and voltage. This relationship involves both a number and an angle. See Table 6.2.

Reactance is the AC generalization of resistance. If this reminds you of vectors, you are right. But it is even better to think in terms of complex numbers (which can be thought of as vectors in two dimensions). If we append a j to the capacitor's reactance[10] to represent the 90° phase shift, we should append a $-j$ to the inductor's reactance. See Table 6.3.

When we write the reactance in this way we call it a *phasor*. I suppose this is because they keep track of the phase shifts between the current and voltage. There is a deep connection between complex numbers and sinusoidal waves, but we will

---

[10] We use the letter j rather than the standard i for complex numbers because it is too easily confused with the I associated with current.

AC Circuits

Component	Symbol	Formula
Resistor	X_R	R
Capacitor	X_C	$(1/2\pi fC)j$
Inductor	X_L	$-(2\pi fL)j$

Table 6.2: AC Reactance Using Phasor Notation

pass over that.[11] These phasors provide a quick way of keeping track of the details of phase shift, etc. Because we have this reactance in the same form for all three components, we can summarize the relationship between voltage and current via a "generalized" Ohm's Law:

$$V = IX \qquad (6.29)$$

where X indicates the reactance of the component.

The advantage of using phasors is when it comes to combining components in a circuit. For any two components, the total reactance is given by

$$\text{Series: } X = X_1 + X_2$$

$$\text{Parallel: } \frac{1}{X} = \frac{1}{X_1} + \frac{1}{X_2}$$

The magnitude of the total reactance is called the *impedance* of the circuit.

Let us now revisit the RC circuit in the context of AC voltage.

---

[11] Euler's equation:
$$\exp(ix) = \cos(x) + i\sin(x)$$

> **AC Circuit – Calculate Impedance**
>
> An AC circuit runs on 100 hertz with a 200 ohms resistor placed in series with a 10 microfarad capacitor. What is the impedance of the circuit? What is the phase shift between the current and voltage?
>
> **Answer:** 260 ohms with a phase angle of 39°

The reactances of the individual components are

$$X_R = 200$$
$$X_C = j/(2\pi)(100)(10 \times 10^{-6}) = 159.16j$$

Since these components are in series, these reactances add.

$$X = X_R + X_C = 200 + 159.16j$$

The impedance is the magnitude of this reactance:

$$Z = \text{mag } \vec{X} = \sqrt{(200)^2 + (159.16)^2} = 255.60$$

and the phase angle is

$$\phi = \text{ang } \vec{X} = \tan^{-1}\frac{\text{Im } X}{\text{Re } X} = \tan^{-1}\frac{(159.16)}{(200)} = 38.51°$$

Suppose now we have all three components in series. The total reactance is

$$X = R + ((2\pi f L) - 1/(2\pi f C))j$$

So the impedance of the circuit is:

$$Z = \sqrt{R^2 + ((2\pi f L) - 1/(2\pi f C))^2}$$

As frequency increases from zero, for a given voltage the magnitude of the current rises from zero (the initial resistance is high because of the capacitor) to a peak at a particular frequency

AC Circuits

then falls again (because the inductor has higher impedance). This resonance peak is the *natural frequency* of the series RCL circuit. It occurs when

$$f_0 = \frac{1}{2\pi\sqrt{LC}} \qquad (6.30)$$

Notice that the resistor does not drive the natural frequency. The resistor does drive the height of the peak.

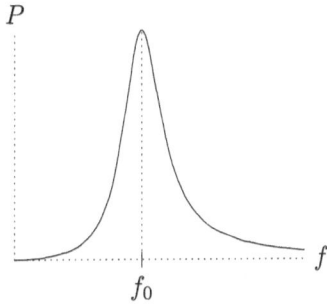

Figure 6.7: Resonance in the RCL circuit

A simple AM radio is based on this RCL circuit. When the circuit is calibrated for a particular frequency only a small voltage is necessary to drive sufficient current to pick up the signal. The electromagnetic waves of this particular frequency are picked up and magnified through the resonance of the RCL circuit. Some capacitors are variable. This is the way to tune the radio to pick up the station you want. All that remains is to amplify the signal and attach a speaker.

AC Circuits – Resonance

Suppose you want to design a circuit to capture radio waves between 60 and 100 kilohertz. With an inductor of 50 millihenries, what range of capacitance do you need?

Answer: Between 140 and 51 picofarads

To get resonance at 60 kilohertz the resonance formula is Equation (6.30). So,

$$(60000) = \frac{1}{2\pi\sqrt{(0.050)(C)}} \Rightarrow C = 1.4072 \times 10^{-10}$$

And for the 100 kHz frequency we have

$$(100000) = \frac{1}{2\pi\sqrt{(0.0050)(C)}} \Rightarrow C = 5.0661 \times 10^{-11}$$

Electromagnetic Waves

In the 1860's Maxwell noticed a mathematical correction was required to the then-known laws of electromagnetism. This collection of laws (with the correction) has become known as Maxwell's laws of electromagnetism. Maxwell noticed that it is possible create a magnetic field that changes in such a way to create a changing electric field that creates a changing magnetic field, etc. In other words, is it possible to create a self-sustaining electromagnetic vibration that feeds itself. Once initiated this electromagnetic wave decouples from the originating charge and flies away.

When Maxwell worked the calculation to determine the speed with which the electromagnetic disturbance moves his result was:

$$c = \sqrt{\frac{1}{\epsilon_0 \mu_0}} = 2.998 \times 10^8 \qquad (6.31)$$

Electromagnetic Waves

EM Radiation	Wavelength	Frequency
Radio	10^3 to 10^{-1}	10^5 to 10^9
Microwaves	10^{-1} to 10^{-4}	10^9 to 10^{11}
Infrared	10^{-3} to 10^{-6}	10^{11} to 10^{13}
Visible	10^{-6} to 10^{-7}	10^{13} to 10^{14}
Ultraviolet	10^{-7} to 10^{-9}	10^{14} to 10^{17}
X-Rays	10^{-9} to 10^{-11}	10^{17} to 10^{18}
Gamma Rays	10^{-11} to 10^{-13}	10^{18} to 10^{21}

Table 6.3: Electromagnetic Spectrum

He discovered that such a wave would move through space at the same speed as light. The implication that light is electromagnetic in origin is obvious. This explanation of the speed of light is one of the great moments in physical science.

After Maxwell described light as electromagnetic waves it became clear that there is a whole range of potential wavelengths and frequencies possible of which visible light is only a small sliver. Table 6.4 describes this electromagnetic spectrum. The table starts with the least energetic ("waves") and ends with the most energetic radiation ("rays"). Note these wavelengths and frequencies are only approximate.

Actually, for the visible spectrum, the wavelength of red light is 7.5×10^{-7} meters and violet light is 3.8×10^{-7} meters. More often this is quoted in nanometers: 750 nm and 380 nm, respectively.

> **Electric Shielding**
>
> If both visible light and radio waves are electromagnetic in origin, explain why it is possible to shield electronic systems from radio waves with a wire mesh that is easy to see through.
>
> **Answer**: Diffraction is proportional to wavelength

The wavelength of radio waves is on the order of 10 meters or so. According to the diffraction formula in Chapter 4, the amount of diffraction depends on the ratio of the wavelength to the slit width. For the wire mesh, the "slits" are on the order of a centimeter or so. This ratio is nearly one thousand, which implies a large amount of diffraction and the energy from the wave is deflected and blocked.

But the wavelength of light is on the order of 10^{-7} meters. The diffraction ratio here is one over 10 million or so. The diffraction is negligible. This is why you can *see* through the mesh without any difficulty.

The electric and magnetic fields in an EM wave are always at right angles to one another. But the orientation of the two together depends on how the wave is generated and can vary erratically. If the waves are purely longitudinal, the wave is said to be *polarized*.

Certain substances are only transparent to EM waves of a particular polarization. We can use them as a filter by passing unpolarized light through them. This is how Polaroid sunglasses work. The Polaroid lens is designed to block glare this way.

What if we place two Polaroid filters on top of one another? If their axes are at an angle, the first will block any light except one polarization and the second will block the component that is parallel its axis. Since this component is proportional to the

Electromagnetic Waves

cosine of the angle between the Polaroid lenses, the intensity of the light will be diminished by the following formula:

$$I = I_0 \cos^2 \theta \qquad (6.32)$$

Polarized Light

When two plates are oriented at 90°, they completely block light. It is interesting that when a third plate is place in between them at an angle, some of the light will be transmitted. Suppose the middle plate is oriented at 45° to the first pair. What is the intensity of light transmitted, as a percent of the incoming light intensity?

Answer: 25%

Since the angle between the first and the middle plate is 45°, the ratio of intensity transmitted is

$$I_1/I_0 = (\cos 45°)^2 = 0.50$$

The ratio of intensity transmitted between the middle and final plate is the same since they are also 45° apart:

$$I_2/I_1 = (\cos 45°)^2 = 0.50$$

The total intensity transfer is the product of these two:

$$I_2/I_0 = (I_2/I_1)(I_1/I_0) = (0.50)(0.50) = 0.25$$

In Newtonian mechanics, waves transport energy from one spot to another. So, we should not be surprised to discover that electromagnetic waves do too. The formula for the amount of electromagnetic energy contained in a small volume V is

$$U = \tfrac{1}{2}(\epsilon_0 E^2 + B^2/\mu_0)V \qquad (6.33)$$

where E and B are the maximum magnitude of the electric and magnetic fields, respectively. If we use Equation (6.31), we can rewrite this as

$$U = \tfrac{1}{2}\epsilon_0(E^2 + c^2 B^2)V$$

For electromagnetic radiation it is a fact that the amount of energy stored in the electric and magnetic fields are the same. Two other formulas follow from this fact:

$$U = (\epsilon_0 E^2)V$$
$$U = (B^2/\mu_0)V$$

Also we have

$$E = cB$$

Remember that these relationships are only true when describing a stand-alone EM wave.

The intensity of a wave of energy is its power transmitted per unit surface area. Or even more fundamentally: the amount of energy transmitted each second through a unit of area. In symbols,

$$I = \frac{E/t}{A}$$

But the EM wave travels though space at a particular speed. In the time frame Δt, the EM wave travels $\Delta x = c\Delta t$ in distance. If we substitute this relation into the previous one we get

$$I = \frac{cE}{A\Delta x} = \frac{cU}{V} = c\epsilon_0 E^2 \qquad (6.34)$$

This shows how the intensity of the radiation is related to the magnitude of its electric field.

EM Radiation – Intensity

The average intensity of sunlight on the earth is approximately 1400 W/m^2. Determine the average magnitude of the electric and magnetic fields from this radiation.

Answer: $E = 730$ volt-meters and $B = 2.4 \times 10^{-6}$ tesla

Electromagnetic Waves

Using Equations (6.33) and (6.34) we should be able to solve this problem. The first step is to use the intensity to figure out the energy content of the radiation with Equation (6.34).

$$(1400) = (3.0 \times 10^8)(U/V) \Rightarrow U/V = 4.6667 \times 10^{-6}$$

The units on this are joules per meter cubed. Now we can use the first variant of Equation (6.33):

$$U/V = (\epsilon_0 E^2)$$
$$\Rightarrow 4.6667 \times 10^{-6} = (8.854 \times 10^{-12})(E)^2$$
$$\Rightarrow E = 726.0$$

The relation $E = cB$ gets us the magnetic field:

$$(726.0) = (3.0 \times 10^8)(B) \Rightarrow B = 2.420 \times 10^{-6}$$

The last topic to cover is how one generates and receives these electromagnetic waves. The key physical mechanism is the acceleration of electric charge. We know that with any electric charge is associated an electric field via Coulomb's Law. If an electric charge moves with a constant velocity, it will generate a magnetic field via Ampere's Law. But this magnetic field is constant. We need a *changing* magnetic field to invoke Faraday's Law to "close the loop" so to speak and generate more electric fields.

So we need a charge that moves with a changing velocity, i.e., accelerating. You shouldn't consider this a proof, but merely a way of justifying the statement that electromagnetic radiation occurs when electric charges accelerate.

Remember that power is energy flow per second. The total rate of radiation from an accelerated charge is given by

$$P = \frac{q^2 a^2}{6\pi\epsilon_0 c^3} \tag{6.35}$$

This is called Larmor's formula. If the particle is moving at relativistic speeds and accelerated in the direction of its motion (e.g., being slowed down quickly), the majority of the radiation is projected forward. This happens in particle accelerators and is called *Bremssrahlung* (braking radiation).

The Classical Atom Is Unstable

In the model of the hydrogen atom due to Niels Bohr, the electron moves around the proton at a speed of 2.2×10^6 m/s in a circle of radius 5.3×10^{-11} meters. Calculate the rate of energy radiated as electromagnetic radiation. How long will it take to radiate away all of the initial kinetic energy of the electron?

Answer: (a) 4.8×10^{-8} J/s, (b) 4.6×10^{-11} seconds

The rate of radiation is given by Larmor's Formula, Equation (6.35), but in order to apply it, we need to know the acceleration of the electron. Since it moves in a circle with uniform speed, the acceleration can be extracted from the formula for centripetal force:

$$a = \frac{v^2}{r} = \frac{(2.2 \times 10^6)^2}{5.3 \times 10^{-11}} = 9.1321 \times 10^{22}$$

This acceleration is huge, but that is because the radius of the circle is so small.

Now we have enough data to use Larmor's Formula:

$$P = \frac{(1.602 \times 10^{-19})^2 (9.1321 \times 10^{22})^2}{(6\pi)(8.854 \times 10^{-12})(3.0 \times 10^8)^3} = 4.750 \times 10^{-8}$$

So this is how many joules of energy is lost in EM radiation each second. This may seem like a small amount, but remember the electron is very light. We want to know how long it takes to

Electromagnetic Waves

radiate its kinetic energy away. That is,

$$KE = \tfrac{1}{2}mv^2$$
$$= \tfrac{1}{2}(9.109 \times 10^{-31})(2.2 \times 10^6)$$
$$= 2.204 \times 10^{-18}$$

Yikes! This is a very small amount of energy. The amount of time it would take to radiate:

$$P = \frac{\Delta E}{\Delta t}$$
$$\Rightarrow (4.750 \times 10^{-8}) = \frac{2.204 \times 10^{-18}}{\Delta t}$$
$$\Rightarrow \Delta t = 4.641 \times 10^{-11}$$

This shows that the Bohr model of the atom is impossible using the laws of classical physics.

One convenient type of accelerated charge is one that oscillates: it stays in one place and it is easy to produce. Easy to produce because all one needs to do is hook up an antenna to an AC circuit. When the electric charge oscillates it produces a dipole radiation pattern. Most of the energy is radiated in the direction of the oscillation and no energy is radiated in the plane perpendicular to the oscillation.

We can capture some of the energy from a passing EM wave with another antenna. This is simply a conductor connected to an amplifier. There are two main types of antenna: linear and loops. The linear antenna is built to react to the passing electric field. The loop antenna is built to react to the passing magnetic field.

Both types of antenna are tuned with an RCL circuit in the amplifier to pick up one particular frequency. The frequency is called the carrier wave. The signal in the wave, which actually contains the message being delivered, is "attached" by slightly modulating either the amplitude of the wave (AM) or the frequency (FM).

This completes our survey of electromagnetism and takes us to the end of the 19th century in physics. In the next chapter we will cover topics discovered in the 20th century.

CHAPTER SEVEN

MODERN PHYSICS

Finally we move solidly into the 20th century. Both relativity and quantum mechanics are traditionally categorized as "modern physics" even though both are now nearly 100 years old. In fact, general relativity (born 1915) is still often glossed over or completely ignored in most introductory texts. The unfortunate reason for these facts is that the time necessary to develop the required mathematics simply cannot be justified at this level.

Therefore the topics in this chapter have traditionally fallen into three groups:

- Special relativity
- Quantum mechanics of the atom
- Nuclear science

Improper Frames

Einstein began the special theory of relativity based upon two postulates:

1. **The Principle of Relativity**

    Physical laws of motion within an inertial frame are the always the same regardless of whether that frame is in motion or at rest.

2. **The Invariance of the Speed of Light**

    Regardless of the motion of the source or observer, the speed of light is always the same: $c = 2.998 \times 10^8$ m/s.

That one can maintain a consistent theory of mechanics which affirms *both* of these apparently contradictory principles is the argument of Einstein's 1905 paper.

The central insight that explains relativity deals with the nature of time.[1] Einstein rejected the Newtonian notion of absolute time. His assault on our common-sense occurs on two fronts. The first has to do with how fast time flows. (60 seconds per minute, right?) This effect is called *time dilation* and obeys the formula

$$\Delta t = \frac{\Delta t_0}{\sqrt{1 - v^2/c^2}} \qquad (7.1)$$

A clock in motion will tick slower than an identical one at rest.

The second assault is called the *relativity of simultaneity*. Suppose two clocks, separated by a certain distance, are synchronized. If these clocks are set into motion, they lose this synchronization. The amount of de-synchronization introduced

---

[1] This is true of both the special and general theories.

Improper Frames

obeys the formula

$$\Delta t = \frac{Lv}{c^2} \tag{7.2}$$

where L is the distance separating the clocks. This means that two events which appear simultaneous to me will not appear simultaneous to you if you are in motion.

These effects are both involved when one measures the length of a moving object. Clearly one needs to measure the front and back *at the same time*. But suppose you are riding along with the object. You will claim that my distance measurement is invalid because my front and back measurements do not look simultaneous to you. After taking these issues into account, one is led to conclude that length measurements are also relative. This effect is called *length contraction* and obeys the formula

$$L = L_0 \sqrt{1 - v^2/c^2} \tag{7.3}$$

A ruler in motion will shrink compared to an identical one at rest.

This $\sqrt{1 - v^2/c^2}$ factor occurs so frequently in relativity that it has acquired a special symbol called the Lorentz factor:

$$\gamma = \frac{1}{\sqrt{1 - v^2/c^2}} \tag{7.4}$$

Using this notation, Equations (7.1) and (7.3) can be shortened to $\Delta t = \gamma \Delta t_0$ and $L = L_0/\gamma$, respectively.

Here is the bottom line. We must accept that the notion "simultaneous" is just as relative as "stationary", that moving rulers shrink, and that moving clocks run slow.

All of these effects highlight the importance of identifying the *proper frame* of measurement. When measuring duration, the proper frame is the one in which the two events occur *in the same place* relative to the frame. In any other frame, this clock will move and appear to tick abnormally slow. The time measured in the proper frame is called the *proper time* between the two events.

When measuring length, the proper frame is the one in which the object measured is at rest. This is usually *not* the same as the proper frame for measuring time. This length is called the *proper length* of the object.

> **Time Dilation**
>
> The muon was first discovered by Carl Anderson in 1936. It can be observed in cosmic radiation on the surface of the earth. The life-time of the muon in the laboratory is observed to be 2.2×10^{-6} seconds. (Technically, this is the half-life of the muon, but let's ignore technicalities for now.) Suppose a muon is created 1.0 kilometers above the surface of the earth with a speed of $0.98c$. Can the muon reach the surface of the earth or not?
>
> **Answer**: Yes

If the particle is moving with $v = 0.98c$, then the Lorentz factor is

$$\gamma = \frac{1}{\sqrt{1 - v^2/c^2}} = \frac{1}{\sqrt{1 - 0.98^2}} = 5.025$$

The *apparent* life-time of the particle is given by Equation (7.1):

$$\Delta t = \gamma \Delta t_0 = (5.025)(2.2 \times 10^{-6}) = 1.106 \times 10^{-5}$$

This is how much time passes between the creation and destruction of the muon as measured on earth. We know the speed, so we can calculate how far the particle gets (as measured on the earth):

$$\Delta x = v \Delta t = (0.98)(3.0 \times 10^8)(1.106 \times 10^{-5}) = 3250$$

Improper Frames

The particle could travel 3.25 km. So the particle does reach the surface of the earth. Without the Lorentz factor, the particle would only travel 0.65 kilometers.

> **Length Contraction**
>
> The earth orbits the sun at a radius of 1.5×10^{11} meters. Calculate the speed of the earth around the sun and determine the resulting length contraction of its diameter (1.3×10^7 meters).
>
> **Answer**: 6.4 centimeters

By definition, speed is distance over duration. The distance involved is the circumference of the earth's orbit. The duration is the length of one year (there are 3.156×10^7 seconds in one year). Thus,

$$v = \frac{2\pi(1.5 \times 10^{11})}{3.156 \times 10^7} = 29863$$

Nearly 30 km/s. The Lorentz factor associated with this speed is

$$\gamma = \frac{1}{\sqrt{1 - \frac{v^2}{c^2}}}$$

$$= \frac{1}{\sqrt{1 - \frac{(2.9863 \times 10^4)^2}{(2.998 \times 10^8)^2}}}$$

$$= 1 + 4.954 \times 10^{-9}$$

Now, this is nearly one. This frequently happens in relativity. A useful result to know is that when $v < 0.01c$, the following approximation is valid.

$$\gamma = \frac{1}{\sqrt{1 - v^2/c^2}} \approx 1 + \tfrac{1}{2}v^2/c^2 \tag{7.5}$$

You can double-check the result above against this equation. The result is the same. The resulting length contraction is

$$L = \frac{1.3 \times 10^7}{1 + 4.954 \times 10^{-9}} = \cdots$$

But this is still pretty unwieldy. Another frequently used approximation[2] is that

$$\frac{1}{1+x} \approx 1 - x$$

when x is small. So $1/\gamma \approx 1 - \frac{1}{2}v^2/c^2$. This allows us to re-write our expression as

$$L = (1.3 \times 10^7)(1 - 4.954 \times 10^{-9})$$
$$= 1.3 \times 10^7 - 0.0644$$

Thus, the apparent reduction in the diameter of the earth is about 6.4 centimeters.

Remember the distance measured by rulers on the earth is in the proper frame, so they measure L_0. The previously calculated length is the *apparent* length of the earth's diameter in motion. This smaller length might be measured by an observer at rest relative to the sun, for example.

Twin Paradox

The twin paradox is perhaps the most celebrated of the many possible paradoxes to consider in special relativity. The story goes something like this.

Consider a pair of twins. Suppose one of these twins stays on earth, and the other becomes an astronaut. The astronaut

---

[2] Both of these approximations are special cases of the "binomial expansion" mentioned in Chapter 1.

Twin Paradox

takes off and travels at a very high speed. According to time dilation, the clocks on her ship tick slower than those on earth. She travels for a while, then turns around to return to earth. When she arrives she is younger than her twin. How can this be?

There are really two levels of weirdness to the story. The first level of weirdness is simply that the two twins are different ages. This is a basic consequence of time dilation. This is not a paradox, *per se*, just an instinctual reaction to the non-Newtonian way that time works. It doesn't feel right for the two twins to have different biological ages—but that is relativity. Weird? Yes. Contradiction? No.

The second level of weirdness is more serious. The origin of the thought lies in the principle of relativity. Suppose we adopt the viewpoint of the traveling twin. She looks back on earth and it appears to be in motion—and moving away at the same speed that she is moving relative to the earth. Why can't she invoke the same time dilation logic? The answer is that she can and she must. She looks back at earth and it is the earth-bound twin that appears to be aging slower! When she returns, she will say that her twin is younger. Which will it be?

Now, this is a true paradox. They both can't be younger than the other. The standard solution is to note that the traveling twin must turn around at some point in order to return to the earth. This acceleration invalidates her frame: it is not truly an inertial frame (one that moves at a constant speed in a constant direction). Therefore her conclusion is false. She is the younger one.

This answer is true, but only tells half of the story. What does the traveling twin actually see? Let's put some numbers into this story to find out.

Twin Paradox – Time Dilation Observed From Earth

An earthbound sibling observes his twin departing from earth at a speed of 0.60c. After 10.0 years, the travelling twin returns having travelled 3.0 light-years (measured from earth) and back. Calculate the proper time the travelling twin was aboard the spaceship.

Answer: 8.0 years

The Lorentz factor for the travelling twin is

$$\gamma = \frac{1}{\sqrt{1-(0.60)^2}} = 1.25$$

The proper time of the travelling twin can be calculated using Equation (7.1) formula:

$$(10) = (1.25)\Delta t_0 \Rightarrow \Delta t_0 = 8.0$$

As the earth-bound twin watches, only 8.0 years pass on the ship.

Here is something to think about. If the travelling twin has only travelled for 8.0 years and she travels the 3.0 light-year distance twice, isn't her speed 0.75c?

No. If the time is measured relative to the ship, so must the distance. According to Equation (7.3), the 3.0 light-year distance is *shorter* for the travelling twin. In fact, the distance from her viewpoint is[3]

$$L = L_0/\gamma = (3.0c)/(1.25) = 2.4c$$

Using this distance, the speed of the ship is still 0.60c.

---

[3] Notice the shortcut that $L_0 = 3.0c$. This works because 3.0 light-years is the distance light travels in 3.0 years. Since $x = vt$, the distance is $L = (c)(3.0)$.

> **Twin Paradox – Time Dilation Observed From Rocket**
>
> From the traveler's vantage point, the earth-bound twin's clocks tick slow. How much time does she observe to pass on earth?
>
> **Answer**: 6.4 years

From the *traveler's* point of view it is the earth that is moving away with a speed of $0.60c$. The Lorentz factor is also 1.25. Since the travelling twin is measuring events on earth, her clocks are not measuring proper time on earth. Using the time dilation formula, the proper time is

$$(8.0) = (1.25)\Delta t_0 \Rightarrow \Delta t_0 = 6.4$$

In other words, as the travelling twin watches, only 6.4 years pass on earth.

> **Twin Paradox – De-synchronization Effect**
>
> During the moment of acceleration 3.0 light-years out, the clocks aboard ship go out of sync with those on earth. Calculate how far ahead the de-synchronization puts the clocks on the ship.
>
> **Answer**: 3.6 years

When the traveler returns either 6.4 years or 10.0 years have passed. Which is it? The truth is that the trip has two legs with a brief period of acceleration in the middle. As the question points out, this acceleration must cover this gap by introducing 3.6 years of perceived time for the traveler. This is the de-synchronization effect summarized in Equation (7.2).

$$\Delta t = \frac{Lv}{c^2} = \frac{(3.0c)(0.60c)}{c^2} = 1.8$$

This number comes from the fact that when she decelerates from 60% the speed of light to zero, there is a 3.0 light-year separation between her and her twin. The time de-synchronization amount is 1.8 years. Then, of course, she accelerates back up to 60% which introduces another 1.8 years of de-synchronization into her clock. The perceived 6.4 years of earth-time plus the 3.6 years of de-synchronization matches the 10.0 years of time seen on earth. Ultimately they both agree that 10.0 years pass on earth.

So, when the traveling twin returns she truly has seen all ten years pass on earth. Paradox resolved.

Relativistic Fusion

It may not be surprising that in special relativity, classical momentum is no longer conserved and that Newton's second law is no longer valid. It needs to be revised into the following form:

$$\gamma F \Delta t = \Delta(\gamma m v) \qquad (7.6)$$

The insertion of the Lorentz factor makes all the difference and this new quantity $\gamma m v$ is conserved. Since for slow speeds this reduces to the standard Newtonian definition, we *replace* the definition of momentum with this new formula as the basic quantity of motion. Thus, relativistic momentum is defined as[4]

$$p = \gamma m v \qquad (7.7)$$

---

[4] A little shop talk. It is a bit of an in-house debate among physics teachers whether the Lorentz factor should be associated with the mass or

Relativistic Fusion 211

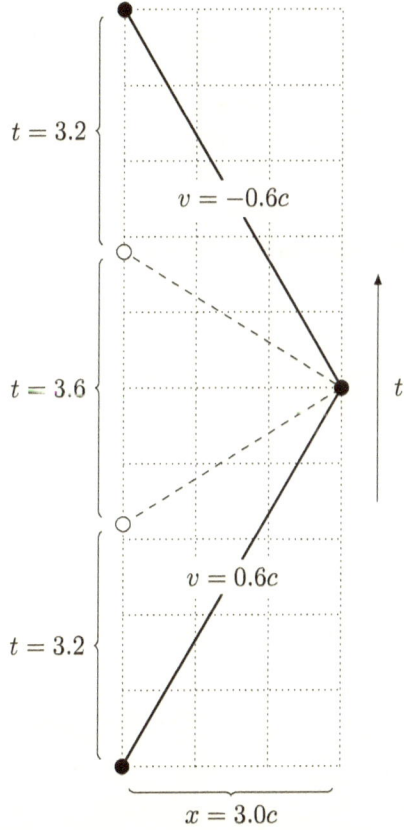

Figure 7.1: The twin paradox from the earth-bound viewpoint

---

the velocity. Currently most authors associate it with the velocity. The implication being that relativity is essentially a change in *kinematics* or the way we measure motion. This also implies that mass is a constant. In the early years of relativity it was common to associate the Lorentz factor with the mass. That is, mass increases with velocity—this is called the relativistic mass. This is (even today) a common reason given for why objects cannot travel faster than light—their mass grows infinitely large the closer they get to light-speed. Interestingly, Feynman (always

> **Relativistic Momentum**
>
> What is the relativistic momentum of an electron (mass = 9.11×10^{-31} kilograms) moving at half the speed of light? At what speed is this momentum doubled?
>
> **Answer**: (a) 1.5779×10^{-22} kg-m/s (b) $0.76c$

First step: find the Lorentz factor.

$$\gamma = \frac{1}{\sqrt{1 - v^2/c^2}} = \frac{1}{\sqrt{1 - (0.5)^2}} = 1.1547$$

Therefore the momentum is

$$\begin{aligned} p &= \gamma m v \\ &= (1.1547)(9.11 \times 10^{-31})(0.5)(3.0 \times 10^8) \\ &= 1.5779 \times 10^{-22} \end{aligned}$$

Now for part (b). Twice this momentum is 3.1558×10^{-22}. We can reverse the momentum definition to extract the velocity.

$$(3.1558 \times 10^{-22}) = (\gamma)(9.11 \times 10^{-31})(v) \Rightarrow \gamma(v/c) = 1.1547$$

It will be easier to solve this for (v/c) than for v. This is sometimes given the symbol β. Thus,

$$\frac{\beta}{\sqrt{1 - \beta^2}} = 1.1547$$
$$\Rightarrow \beta^2 = (1.3333)(1 - \beta^2)$$
$$\Rightarrow (2.3333)\beta^2 = (1.3333)$$
$$\Rightarrow \beta = 0.7559$$

So, the momentum is doubled when the speed increases from 50% of the speed of light to 76% of the speed of light.

---

the contrarian) states that *all* of special relativity follows from the idea of relativistic mass.

Relativistic Fusion

The modification of Newton's Second Law in Equation (7.6) and the re-definition of momentum in Equation (7.7) propagates through the rest of mechanics. In particular, the idea of energy must be revised. We still continue to define work as force multiplied by displacement and kinetic energy as the work required to bring an object up to speed from rest. The resulting formula for kinetic energy is

$$KE = (\gamma - 1)mc^2 \qquad (7.8)$$

This doesn't look like the classical formula for kinetic energy until one rewrites the Lorentz factor using Equation (7.5). Thus,

$$KE \approx (1 + \tfrac{1}{2}v^2/c^2 - 1)mc^2 = \tfrac{1}{2}mv^2$$

This shows that Equation (7.8) does in fact reduce to the classical formula with which we are familiar.

Remember that momentum is conserved in all collisions but kinetic energy is only conserved in elastic collisions. However, there is an interesting consequence of Einstein's new definition of momentum. It motivates re-defining energy such that it is conserved in any collision: elastic or not! One way to show this is by starting with the Lorentz factor.

$$\gamma = \frac{1}{\sqrt{1 - v^2/c^2}}$$

Square both sides and cross-multiply the fraction to yield

$$\gamma^2(1 - v^2/c^2) = 1$$

Then distribute the γ^2 and multiply both sides by c^2 to get

$$\gamma^2 c^2 - \gamma^2 v^2 = c^2$$

The γv term should remind you of the new definition of momentum, $p = \gamma m v$. If we multiply both sides of this equation by m^2 (rest mass squared), we have

$$(\gamma mc)^2 - p^2 = (mc)^2$$

The quantity on the right side is the rest mass of the particle times the speed of light which is just a number, a constant. Since the momentum p is conserved, so must the first term (γmc) on the left. Otherwise they could not always add to the same number. But what is this thing?

There are three ways to describe it. The first is as the "time-component" of momentum. That is most consistent with how we just discovered it. Though not unheard of, it is not the usual description.

A second way is to notice that the c is just a constant also, so really we are talking about the product γm as the thing being conserved. You would say then that we have proven the conservation of *relativistic mass*. This also has the advantage of being consistent with the Newtonian collisions rules: momentum and mass are conserved in any collision.

However, the third way to describe this quantity is the most common. Instead of dividing by c, multiply by c. The quantity γmc^2 has the units of energy. This motivates *redefining* energy as given by this formula. When we do that the relationship above becomes

$$E^2 - p^2 c^2 = m^2 c^4 \tag{7.9}$$

where

$$E = \gamma mc^2 \tag{7.10}$$

Of course when $v = 0$, the Lorentz factor becomes one, so we have

$$E_0 = mc^2 \tag{7.11}$$

perhaps the most famous formula in physics. This states that even when a particle is at rest is has energy. This is called its *rest energy*.

Relativistic Fusion

> **Relativistic Fusion**
>
> Two identical particles with rest mass of 4.0 units collide. They both have a speed of $0.6c$ before the collision. The collision is completely inelastic so the composite particle is at rest. What is its mass?
>
> **Answer:** 10 units

As usual, we begin by calculating the Lorentz factor for the incoming particles.

$$\gamma = \frac{1}{\sqrt{1-(0.6)^2}} = 1.25$$

The momentum of each particle is given by $p = \gamma m v$, but they both cancel each other out. We also now know that the relativistic energy is conserved (even in this inelastic collision). The energy of each incoming particle is

$$E = \gamma m c^2 = (1.25)(4.0)(c^2) = (5.0)(c^2)$$

The momenta cancel, but the energies add. Therefore the composite particle must have an energy level of $10c^2$ units. However, the composite is not moving, so the Lorentz factor is one. In other words, this is its rest energy. Diving by c^2 yields a final rest mass of 10 units.

This problem illustrates that the final rest energy is more than the combination of the rest energies of the incoming particles. We know that in an inelastic collision the incoming kinetic energy becomes hidden in internal heat energy. But the internal energy is a summary of the kinetic and potential energies of the molecules that make up the system. In relativity, all of these forms of energy contribute mass to the system.

Photo-Electric Effect

Einstein did not receive the 1921 Nobel Prize for his work in relativity, rather "for his services to theoretical physics, and especially for his discovery of the law of the photoelectric effect."[5] The theory of relativity was a little too radical for the committee to grant the prize.

The significance of the photo-electric effect is that it offers evidence of the "corpuscular nature" of light. In other words, light acts like a particle rather than a wave. This is significant because going back to the early 1800's physicists had assumed that light acts like a wave. The truth is that light has both particle-like and wave-like characteristics. The reconciliation of these two contrary models of matter is the subject of quantum mechanics.

How does the photo-electric effect work? If one shines light on a metallic surface under the right circumstances, electrons will fly off the metal. These electrons can be captured and measured as current.

The interesting thing is the circumstances required to generate the photo-electric effect. It is found that the effect is frequency dependent. The formula is

$$KE_{\max} = hf - W \qquad (7.12)$$

where $KE_{\max}$ represents the maximum kinetic energy of the ejected electrons and W is a constant determined by the metal. It is called the "work function" of the metal.

One might expect that by increasing the intensity of the light, the kinetic energy absorbed by the electrons would increase. This expectation is based on wave theory. For any oscillation,

---

[5] http://nobelprize.org/nobel_prizes/physics/laureates/1921/

Photo-Electric Effect

the frequency is controlled by the nature of the system.[6] The energy of the oscillation is manifested in the amplitude of the motion. For electromagnetic radiation, the amplitude of the wave is related to the size of the electric field. One expects the electrons to be more excited by a high intensity wave than a lesser intense one.

But this doesn't happen. When you increase the intensity of the light, the kinetic energy of the electrons is unchanged. You do get more of them, but each individual electron does not possess more kinetic energy. Only by changing the frequency of the light can you change the kinetic energy of the electrons. But, in electromagnetic theory, the frequency is unrelated to energy.

Einstein's solution to this conundrum was simple yet bold. He rejected the electromagnetic wave model for light and reverted back to the Newtonian "corpuscular" model of light. In modern lingo: the photon. Planck had already discovered that when one assumes that electromagnetic energy is constrained to energy levels of

$$E = hf \quad (7.13)$$

one can avoid something called the "ultraviolet catastrophe".[7] But Einstein pictured this constraint as something more like a light-particle. The formula for the photo-electric effect, Equations (7.12), is simply the conservation of energy applied to these photons.

---

[6] For a spring, it is the spring constant and the mass. For a pendulum, the acceleration due to gravity and the length of the pendulum.

[7] This is a prediction from classical physics that when an object is heated it will radiate an infinite amount of electromagnetic energy in the ultraviolet end of the spectrum. This is clearly wrong because hot objects tend to glow red not blue!

> **Photo-electric Effect**
>
> The work function for sodium is 2.36 electron-volts. When illuminated with 350 nanometer light, what is the maximum speed of the electrons ejected from the sodium?
>
> **Answer**: 6.46×10^5 m/s

The frequency of this light is

$$f = \frac{c}{\lambda} = \frac{3.0 \times 10^8}{350 \times 10^{-9}} = 8.5714 \times 10^{14}$$

So the energy of each photon of light is

$$\begin{aligned} E &= hf \\ &= (6.626 \times 10^{-34})(8.5714 \times 10^{14}) \\ &= 5.6794 \times 10^{-19} \end{aligned}$$

The conversion from joules to electron-volts yields

$$E = (5.6794 \times 10^{-19} \text{ J}) \times \left(\frac{1 \text{ eV}}{1.602 \times 10^{-19} \text{ J}}\right)$$
$$= 3.5452 \text{ eV}$$

It takes 2.36 electron-volts to ionize the sodium, so the electrons have at most 1.1852 electron-volts of kinetic energy. In joules, we have

$$KE_{\max} = (1.1852 \text{ eV}) \times \left(\frac{1.602 \times 10^{-19} \text{ J}}{1 \text{ eV}}\right)$$
$$= 1.8987 \times 10^{-19}$$

We can extract the speed from the kinetic energy using

$$\begin{aligned} KE &= \tfrac{1}{2}mv^2 \\ \Rightarrow (1.8987 \times 10^{-19}) &= \tfrac{1}{2}(9.109 \times 10^{-31})(v)^2 \\ \Rightarrow v &= 6.4566 \times 10^5 \end{aligned}$$

At the turn of the 20th century, the evidence for the particle-like nature of light began to pile up. However, the old evidence for the wave nature of light is still valid (e.g., interference, diffraction, superposition). In some cases light acts like a particle and in other cases it acts like a wave. The confusion was ultimately resolved with quantum mechanics in the 1930's. But before this resolution, the confusion got worse before it got better.

Spectroscopy

When Rutherford discovered that the bulk of the atom is contained in a small positively charged nucleus, a planetary-type model of the atom was obvious. Unfortunately, a simple electromagnetic calculation using Larmor's Formula shows that this model is unstable. The electron would spiral into the nucleus shedding its kinetic energy as electromagnetic radiation in nanoseconds.

Bohr's model of the atom is not so much a solution to the problem as it is a kind of stop-gap. He simply assumed that the electron orbits like a planet without radiating. But it does so in such a way that its angular momentum is "quantized" by the formula

$$L = nh/2\pi \tag{7.14}$$

Something like Planck did by assuming that radiation is quantized in Equation (7.13). This assumption implies that the electron only orbits at certain radii. It follows that the energy levels of the electron are

$$E = -(13.6 \text{ eV})\frac{Z^2}{n^2} \tag{7.15}$$

where Z is the number of protons in the nucleus. Bohr also postulated that when the atom radiates (or absorbs) energy, it does so between these levels.

> **Simple Spectroscopy**
>
> What is the wavelength of light that the hydrogen atom will emit when an electron falls from the $n = 2$ orbit to the $n = 1$ orbit? What is the wavelength for the same transition in an helium atom?
>
> **Answer:** Hydrogen: 122 nanometers; Helium: 30.4 nanometers

The energy for these levels in hydrogen ($Z = 1$) are:

$$E_1 = (13.6 \text{ eV})\frac{(1)^2}{(1)^2} = 13.6 \text{ eV}$$

$$E_2 = (13.6 \text{ eV})\frac{(1)^2}{(2)^2} = 3.4 \text{ eV}$$

The difference between them is 10.2 electron-volts. The corresponding frequency is

$$E = hf$$
$$\Rightarrow (10.2 \text{ eV}) \times \left(\frac{1.602 \times 10^{-19} \text{ J}}{1 \text{ eV}}\right) = (6.626 \times 10^{-34})(f)$$
$$\Rightarrow f = 2.4661 \times 10^{15}$$

So, its wavelength is

$$\lambda = \frac{c}{f} = \frac{3.0 \times 10^8}{2.4661 \times 10^{15}} = 1.2165 \times 10^{-7}$$

or, 122 nanometers.

A similar calculation works for helium ($Z = 2$):

$$E_1 = (13.6 \text{ eV})\frac{(2)^2}{(1)^2} = 54.4 \text{ eV}$$

$$E_2 = (13.6 \text{ eV})\frac{(2)^2}{(2)^2} = 13.6 \text{ eV}$$

De Broglie Waves

The difference between them is 40.8 electron-volts. The corresponding frequency is

$$E = hf$$
$$\Rightarrow (40.8 \text{ eV}) \times \left(\frac{1.602 \times 10^{-19} \text{ J}}{1 \text{ eV}}\right) = (6.626 \times 10^{-34})(f)$$
$$\Rightarrow f = 9.8644 \times 10^{15}$$

So, its wavelength is

$$\lambda = \frac{c}{f} = \frac{3.0 \times 10^8}{9.8644 \times 10^{15}} = 3.0412 \times 10^{-8}$$

or, 30.4 nanometers, which is one-quarter of the value for hydrogen.

De Broglie Waves

Bohr's model is successful at explaining the observed spectrographic pattern of the elements, but one is left wondering *why* the electrons are constrained in this way. De Broglie conjectured that the electron is associated with a wavelength given by the formula

$$\lambda = h/p \quad (7.16)$$

In order for this "electron wave" to maintain a circular orbit, it must be a standing wave. Thus, the circumference of the orbit must an integer multiple of half-wavelengths. This "explains" Bohr's postulate.

De Broglie's hypothesis of *matter waves* has subsequently been observed directly in experiments involving electron diffraction, neutron interference, etc. It seems that in the microscopic realm, matter and energy display both wave- and particle-like properties.

> **De Broglie Wavelength**
>
> An electron diffraction experiment requires electrons with a wavelength of 10^{-10} meters. What voltage potential is required to accelerate an electron with sufficient momentum to have this wavelength?
>
> **Answer**: 150 volts

The momentum required is given by Equation (7.16).

$$\begin{aligned} p &= h/\lambda \\ &= (6.626 \times 10^{-34})/(1 \times 10^{-10}) \\ &= 6.626 \times 10^{-24} \end{aligned}$$

It is unclear whether we need to be careful with relativity or not. Usually the simplest solution is to cross your fingers and try it with the Newtonian formulas and confirm that the final answer is in a non-relativistic range. The classical formula for kinetic energy is

$$KE = \frac{p^2}{2m} = \frac{6.626 \times 10^{-24})^2}{(2)(9.109 \times 10^{-31})} = 2.4099 \times 10^{-17}$$

After converting to electron-volts we have:

$$KE = (2.4099 \times 10^{-17} \text{ J}) \times \left(\frac{1 \text{ eV}}{1.602 \times 10^{-19} \text{ J}}\right)$$
$$= 150.4 \text{ eV}$$

Since we are talking about an electron in this problem, the voltage required to accelerate the particle up to this energy is simply 150 volts. This puts us solidly in the non-relativistic range, so we can stop here.

One interesting consequence of these ideas is the Heisenberg's uncertainty principle. It states that the position and the mo-

De Broglie Waves

mentum of a quantum particle cannot be determined with arbitrary precision. The relationship is

$$(\Delta x)(\Delta p) \geq h/4\pi \qquad (7.17)$$

Given the relationship between relativistic momentum and energy, does it surprise you that the measurement precision in energy and time is also constrained? The relationship is

$$(\Delta E)(\Delta t) \geq h/4\pi \qquad (7.18)$$

> ### Uncertainty Principle – Oyster Diagram
>
> In quantum electrodynamics, it is possible for the vacuum to "borrow" energy to create an electron-positron pair for a short period of time before they recombine back to the original vacuum state. This "bubbling" of the vacuum contributes to what is known as "vacuum energy". Estimate the period of time the particle pair exists using the energy-time version of the uncertainty principle.
>
> **Answer:** 3.2×10^{-19} seconds

The energy that is being "borrowed" from the vacuum is the rest energy of the electron-positron pair. That is:

$$\begin{aligned} E &= 2mc^2 \\ &= (2)(9.109 \times 10^{-34})(3.0 \times 10^8)^2 \\ &= 1.6396 \times 10^{-16} \end{aligned}$$

According to Equation (7.18), the time interval associated with this change in energy is:

$$\begin{aligned} (1.6396 \times 10^{-16})(\Delta t) &\geq (6.626 \times 10^{-34})/4\pi \\ \Rightarrow \Delta t &= 3.216 \times 10^{-19} \end{aligned}$$

An immeasurably small period of time (so far).

Name	Symbol	Values
Energy level	n	$1, 2, 3, \ldots$
Angular momentum	ℓ	$0, 1, \ldots, (n-1)$
Magnetic number	m_ℓ	$\pm \ell, \pm(\ell-1), \ldots, 0$
Quantum spin	m_s	$\pm \frac{1}{2}$

Table 7.1: Atomic Quantum Numbers

Atomic Orbitals

Of course, the electron is not a string, but moves in three dimensions. So one really ought to model the electron as a standing three-dimensional wave (something like a sound wave from a point source). This picture obliterates the old Bohr model. It is replaced with the idea of an electron *orbital*. These orbitals are characterized by the four "quantum numbers" listed in Table 7.1.[8]

These numbers correspond to the different *shape* of each standing wave. Each orbital interacts with the orbitals in other atoms in a particular way. This is what we call chemistry.

In general, the electrons fill these orbitals from least to most energetic. The Pauli *exclusion principle* guarantees that no two electrons occupy the same quantum state.

Thus, as we add electrons to the system, they begin to "fill up" the electron shells. Two can sit in the $n = 1$ state since they may have opposing spins. These two quantum states are distinct but have the same energy. We say they are *degenerate* states. This means there is a two-fold degeneracy in the first

---

[8] By the way, the reason for *four* numbers is the number of dimensions in space-time.

Atomic Orbitals

energy level. So two electrons can occupy the $n = 1$ state before the exclusion principle forces the next one into the $n = 2$ state. Because they have different magnetic moments, the energy levels of the degenerate orbitals can be manipulated and split by applying a magnetic field. This is why the third quantum number is called "magnetic".

By convention (from the study of spectroscopy), orbitals with $\ell = 0, 1, 2, 3$ are designated with the letters s, p, d, and f, respectively. The order in which the orbitals are filled is listed in Table 7.2.

Electronic Configuration

A carbon atom has 12 electrons. What is its electronic configuration in the ground state?

Answer: $1s^2\ 2s^2\ 1p^6\ 3s^2$

This is pretty simple. We look up Table 7.2 to see how the orbitals fill and write down the orbitals as they fill. In our case we have

n	ℓ	Label	Degeneracy	Cumulative
1	0	1s	2	2
2	0	2s	2	4
2	1	1p	6	10
3	0	3s	2	12

So the electronic configuration is written as $1s^2\ 2s^2\ 1p^6\ 3s^2$. The fact that carbon fills that 3s sub-shell is one of the reasons it is a little more stable than its periodic table neighbors.

Not all quantum particles obey the Pauli exclusion principle. A particle that does is called a *fermion*, one that doesn't is

n	ℓ	Label	Degeneracy	Cumulative
1	0	1s	2	2
2	0	2s	2	4
2	1	1p	6	10
3	0	3s	2	12
3	1	3p	6	18
4	0	4s	2	20
3	2	3d	10	30
4	1	4p	6	36
5	0	5s	2	38
4	2	5p	6	54
6	0	6s	2	56
4	3	4f	14	70
5	2	5d	10	80
6	1	6p	6	86
7	0	7s	2	88
5	3	5f	14	102
6	2	6d	10	112

Table 7.2: Atomic Orbitals Sorted By Energy

called a *boson*. All quantum particles are either one or the other. If fermions are exclusive, bosons are gregarious. That is, they tend to aggregate into the *same* quantum state.

Photons are examples of bosons. This helps explain how lasers create light with such high coherency. In fact, when two electrons are coupled, the pair can also act like a boson. This helps to explain the covalent bond in chemistry, ferromagnetism, and superconductivity in solid state physics.

There is plenty more to talk about, but that's as far as we will go into quantum mechanics. We now move on to the third major subject of modern physics: nuclear science.

Nuclear Energy

By the 1930's, the structure of the atom was pretty well understood. These investigations culminated in the 1950's with a comprehensive quantum theory of electromagnetism: QED. To date this is the most precisely tested and successful scientific theory known to man.[9]

When it comes to the nucleus of the atom, it is a very different matter. A few basic facts are clear. The nucleus is a composite system with at least two elements: protons and neutrons. Also, there must be some form of short-range nuclear force that holds it all together, counter-balancing the electrostatic repulsion of the charged protons. Actually, there are *two* forces that govern the sub-atomic world. They are cleverly called the strong and weak nuclear forces.[10] It is the strong force that holds the nucleus together.

---

[9] This is no exaggeration. Some calculations have been confirmed to an accuracy of 1 part per billion.

[10] You will quickly discover that the creativity of the physicists to name these new nuclear phenomena was severely strained.

The delicate balancing act between electrostatic repulsion and the attractive strong nuclear force is responsible for both nuclear fission and fusion.

Fission occurs when the nucleus falls apart. Due to the short-range nature of the strong force, once the nucleus reaches a certain size, its attractive tension can no longer resist the repulsion between the protons and the nucleus becomes unstable. The uncharged neutrons do act as a stabilizing influence (increasing the nuclear attraction without contributing to the electrostatic repulsion), but they can only help so much. Once the nucleus has about 200 nucleons (that is, protons or neutrons), the nucleus has a tendency to fall apart. Lead-208 has the largest stable nucleus with 82 protons and 126 neutrons. The largest naturally occurring element is uranium-238.

Because of these competing forces, each nucleus is wound up like a bound spring. If one can be broken, the electrostatic potential energy can be released as kinetic energy. Nuclear fission is essentially a completely inelastic collision in reverse. According to relativity, this bound up energy contributes to the mass of the nucleus as a consequence. The nucleus is actually more massive than the sum of its parts. This gives us a way to estimate the amount of energy released in nuclear fission.

> **Nuclear Fission**
>
> Suppose a neutron strikes the nucleus of uranium-235 and the combination splits into barium-141, krypton-92, and three neutrons. How much kinetic energy is released in the reaction?
>
> **Answer:** 2.78×10^{-11} joules

The masses of these particles are

Nuclear Energy

Name	Symbol	Atomic Mass
Uranium-235	$^{235}_{92}\text{U}$	235.043930
Barium-141	$^{141}_{56}\text{Ba}$	140.914411
Krypton-92	$^{92}_{36}\text{Kr}$	91.926156
Neutron	$^{1}_{0}\text{n}$	1.008665

It is typical to quote these masses relative to the *atomic mass unit*, which is defined such that carbon-12 is exactly 12 atomic mass units (the symbol is simply "u"). The conversion to SI units is

$$1 \text{ u} = 1.6605 \times 10^{-27} \text{ J}$$

The total mass of the incoming particles in this problem is

$$m_n + m_U = (1.008665) + (235.043930)$$
$$= 236.052595$$

And the total mass of the outgoing particles is

$$m_{\text{Ba}} + m_{\text{Kr}} + 3m_{\text{n}} = (140.914411)$$
$$+ (91.926156)$$
$$+ 3(1.008665)$$
$$= 235.866562$$

The *mass defect* is the difference between these two. Since the mass of the outgoing particles is less than the incoming particles, the remaining mass-energy is released as kinetic energy. Thus,

$$\Delta m = (236.052595) - (235.866562)$$
$$= 0.186033$$

We can convert the atomic mass units into kilograms, then convert this mass using $E = mc^2$. But this is such a common calculation that the conversion is simpler if we just use the formula

$$(1 \text{ u})(c^2) = 1.492 \times 10^{-10} \text{ J}$$

Therefore, the total kinetic energy released is

$$KE = (0.186033)(1.492 \times 10^{-10}) = 2.7756 \times 10^{-11}$$

This may not seem like a lot, but this is for *each* nuclear reaction. If we have something like 10^{23} of these reactions, the amount of energy released can be quite explosive. (Pun intended.)

Fusion is another way nuclear energy is released. This is the opposite process: two nuclei collide and combine to create a larger single nucleus. The reason this releases energy is that the nuclear force is much stronger that the electrostatic force. It is possible for the two nuclei to get close enough together for the attractive nuclear force to overcome the electrostatic repulsion. If so, the two nuclei roll into the potential created by the nuclear force and pick up kinetic energy. Once the composite nucleus stabilizes, this residual energy is released in the form of radiation.[11]

---

[11] Not just gamma radiation. It is possible that only parts of the incoming nuclei are in the composite. The remaining pieces fly off like nuclear shrapnel.

Nuclear Energy

> **Nuclear Fusion**
>
> One of the atomic processes to create nuclear fusion involves two isotopes of hydrogen. When a deuterium nucleus combines with a tritium nucleus, they fuse into a helium nucleus and an extra neutron. Most of the surplus energy is carried away by kinetic energy in the extra neutron. Assume all of the energy is carried away by the neutron. How fast is it moving as a percent of the speed of light?
>
> **Answer:** 19%

The masses of these particles are

Name	Symbol	Atomic Mass
Deuterium	$^2_1 H$	2.014102
Tritium	$^3_1 H$	3.016050
Helium	$^4_2 He$	4.002603
Neutron	$^1_0 n$	1.008665

The total mass of the initial particles (in atomic mass units) is

$$(2.014102) + (3.016050) = 5.030152$$

and the total mass of the final particles is

$$(4.002603) + (1.008665) = 5.011268$$

Since there is more mass initially, the reaction must shed energy in the decay products. The mass-energy that is carried away by the kinetic energy of the neutron is

$$(5.030152) - (5.011268) = 0.018884$$

but we must multiply this result by c^2 in order to convert from units of mass to units of energy.

Since the kinetic energy is given by
$$KE = (\gamma - 1)mc^2$$
we can calculate γ then extract v. Thus,
$$(0.018884)(c^2) = (\gamma - 1)(1.008665)(c^2)$$
$$\Rightarrow \gamma = 1.01872$$
$$\Rightarrow v/c = 0.1908$$
So, the neutron is moving at just over 19% of the speed of light.

Radioactive Half-Life

As I mentioned previously there are actually two short range sub-atomic forces. So far we have only spoken of the strong one. The strong nuclear force is sufficient to explain nearly all nuclear phenomena save one: the instability of the neutron. As long as the neutron is surrounded by other nucleons (protons or neutrons), the particle is stable. But if separated, the neutron has a half-life of about 10 minutes. This neutron decay is responsible for beta radiation.

If you start with 1000 separated neutrons, after 10 minutes you will only have 500 left. After another 10 minutes you will have half of that: 250. After another 10 minutes (30 minutes total), you will only have 125. And so on.

When the neutron decays, it releases a proton and an electron. However, careful observations in the 1930's showed that these two particles were insufficient to account for the conservation of energy and momentum. Either the neutron routinely violates these laws or an unobserved particle is involved. This particle is now called an anti-neutrino.[12]

---

[12] This particle was originally simply called a neutrino, Italian for "little neutral one". The "anti" is there because modern quantum field theory requires that this be particle of anti-matter.

Radioactive Half-Life

It's not really correct to think of the neutron as a composite of a proton, electron, and anti-neutrino held together by a weak nuclear force in some sort of sub-atomic dance. It is more proper to think of the neutron as actually transforming into the combination in a kind of material annihilation/creation process. We will return to these ideas in the last section of this chapter.

Consider carbon. Each carbon nucleus has six protons. (Remember: the electrons determine the chemical nature of the element and the numbers of protons in the nucleus control the number of surrounding electrons. The neutrons are there just to help hold the protons together). The most abundant isotope of carbon has six neutrons: carbon-12.

Another relatively abundant form of carbon has eight neutrons: carbon-14. However, this nucleus is just a bit too large to be stable. The neutrons on the periphery are just beyond the stabilizing influence of the other nucleons. As a consequence, carbon-14 is radioactive. The half-life of carbon-14 is about 5730 years.

Carbon-14 is created in the atmosphere by cosmic radiation. So even though the carbon-14 is constantly dwindling through radioactive decay, it is being supplied by this cosmic radiation. The amount of carbon-14 in the world represents an equilibrium between the continual creation and natural decay processes.

And this is why radiocarbon dating works. Since living creatures use carbon, the ratio of carbon-14 to carbon-12 matches the equilibrium ratio in the world. However, when a creature dies it no longer ingests this natural carbon equilibrium. The carbon-14 in its body decays with a half-life of 5730 years. Each neutron decay acts like a ticking clock measuring the time since the death of the creature. By measuring the ratio between the carbon-14 and the carbon-12 in a sample, one can get an estimate of the age of the sample.

> **Radiocarbon Dating**
>
> An Egyptian wooden box is analyzed using radiocarbon dating. The ratio of carbon-14 to carbon-12 in the specimen is measured to be between 0.45 parts per trillion and 0.55 parts per trillion. Assuming the current ratio is 1.0 parts per trillion, give an estimate the age of the box.
>
> **Answer:** 5800 ± 800 yr

The key formula here is that the measured ratio is related to the current ratio according to

$$N/N_0 = (0.5)^{t/h}$$

where h is the half-life of carbon-14: 5730 years. Using the low ratio we get

$$0.45 = (0.5)^{t_1/h}$$
$$\Rightarrow \ln(0.45) = \ln((0.5)^{t_1/5730})$$
$$\Rightarrow \ln(0.45) = (t_1/5730)\ln(0.5)$$
$$\Rightarrow -0.7985 = (t_1/5730)(-0.6931)$$
$$\Rightarrow t_1 = 6600$$

And the high ratio gives

$$0.55 = (0.5)^{t_2/h}$$
$$\Rightarrow \ln(0.55) = \ln((0.5)^{t_2/5730})$$
$$\Rightarrow \ln(0.55) = (t_2/5730)\ln(0.5)$$
$$\Rightarrow -0.5978 = (t_2/5730)(-0.6931)$$
$$\Rightarrow t_2 = 4942$$

In order to quote these together we calculate the average and the half-range between them:

$$(t_1 + t_2)/2 = 5771$$

and

$$(t_1 - t_2)/2 = 829$$

The final estimate is therefore

$$t = 5771 \pm 829$$

But, really, since the error range is so large, it is not appropriate to quote all these digits as though they are significant. The best answer is

$$t = 5800 \pm 800$$

This same trick can be used with other elements and isotopes. For example, the ratio of uranium-235 to uranium-238 is often used to date rocks because the radioactive half-lives are on the order of 0.7 and 4.5 billion years, respectively.

Quarks And The Standard Model

We will finish this chapter with a discussion of the *Standard Model* of elementary particles. The Standard Model uses the ideas of quantum electrodynamics and applies them to nuclear phenomena. It is therefore necessary for us to discuss some ideas involved in quantum electrodynamics (QED).

In QED, the electromagnetic force is mediated by photons of light. That is, two charged particles exchange photons. One charged particle emits a photon and this photon carries away momentum and energy. The momentum and energy is then captured by a second charged particle. Based on the quantum rules of QED, this exchange is what causes attraction or repulsion depending on the nature of the particles, their separation in space, etc.

From this perspective, there are no forces: just particles exchanging other particles. The exchange particles are all bosons that are created or absorbed by fermions. The bosons correspond to forces and the fermions correspond to matter. The mathematics is extremely complicated, but this is the essence of the theory.

When it comes to applying these ideas to nuclear matter, the first step is to identify the relevant bosons and fermions. The high-energy collision experiments in the 1950's and 1960's produced a surprising diversity of particles previously unknown. The implicit order underlying this "particle zoo" is what the Standard Model describes.

A first guess is that protons and neutrons must exchange some particle. At first it appeared that a newly discovered particle called the pion would do the trick, but eventually it was realized that this scheme could not work. The solution lay at a more fundamental level.

The final solution was the conjecture that the proton and neutron are actually composed of smaller particles called *quarks*. Two quarks are necessary: the "up" and "down" quarks. Don't get confused: there is no real "up-ness" or "down-ness" to these particles. It's merely that up and down are opposites: similar but different.

Each proton and neutron is made of a triplet of quarks. The proton is made of two up quarks and one down quark while the neutron is made of one up quark and two down quarks. In symbols,

$$p = uud$$
$$n = udd$$

In order for this scheme to work, the up quark must have a charge of $+2/3$ and the down quark must have a charge of $-1/3$.

The quarks are attracted to one another via the strong force

Quarks And The Standard Model

which is mediated by a new type of boson: the *gluon*.[13] The gluon holds the same position in the strong force as the photon does in electromagnetism. These gluons mediate a new type of force with a new type of "charge" called *color*. Each quark carries this property and the gluons transmit this quark-color back and forth. This produces the attractive forces that hold the proton and neutron together. This also accounts for the residual attraction between protons and neutrons that hold the nucleus together.

> **Quark Content of Pion**
>
> The pion was initially considered (incorrectly) as the particle that mediates the strong interaction between a pair of nucleons. This means, for example, a proton can convert into a neutron by absorbing a negative pion. Knowing that the quark content of the proton is uud and the neutron is udd, what must the quark content of the negative pion be?
>
> **Answer:** $\bar{u}d$

The negative pion (π^-) must change the quarks triplet from

$$uud \to udd$$

Since both the proton and neutron have one u quark and one d quark, the effect of the pion must be to convert a u quark into a d quark. Said differently, it must destroy the u and leave the d. The thing that annihilates a particle is its anti-particle. In this case the anti-u quark, $\bar{u}$. Therefore the negative pion must be composed of two quarks: an anti-up quark and a down quark. In symbols,

$$\pi^- = \bar{u}d$$

Notice that the charge of the $\bar{u}$ quark is $-2/3$, so the total charge is -1.

---

[13] I'm not making this up. This is really its name.

This example shows that quarks can also combine in pairs. When they do it is always a quark with another anti-quark. These combinations are called *mesons*. The triplet combinations (like protons and neutrons) are called *baryons*. Quarks can only exist in one of these two combinations. In particular, they cannot exist in isolation.

The quantum chromodynamics (QCD) describing the strong force would be sufficient if it were not for the decay of the neutron. This decay implies some tendency for the d quark to transform into the u quark spontaneously.[14] This suggests that truly all of these quarks are fundamentally the same, but wear different clothes depending on whether they are up or down. This property of the quark is called its *flavor*.[15]

In a similar way as in QED and QCD, one imagines a boson that corresponds to this transformation. In fact it also mediates the transformation of the neutrino into an electron. There are details I simply don't have time to discuss (e.g., one really needs three new bosons called the W^+, W^-, and the Z in order to make the scheme work), but these new bosons explain why these four particles are involved in beta decay.

This is the Standard Model of fundamental physics. In the end, we have four fundamental types of fermions that form the entire material world:

- Up Quarks
- Down Quarks
- Electrons
- Neutrinos

And collection of bosons mediating the interactions between

---

[14] Without the emission of a negative pion, for example.

[15] Seriously. I'm not making this up.

Quarks And The Standard Model

them:

- Photons

- Gluons

- Weak Bosons ($W^{\pm}$, Z)

Inexplicably, the story does not end here. There are more particles in nature and in the laboratory than these. For example, the muon is in every respect identical to the electron except it is much more massive. After its discovery in 1936, Rabi famously quipped, "Who ordered that?"

Amazingly, the structure of the Standard Model need not change: we simply add two more "generations" of fermions:

Quarks	Leptons
Up (u)	Electron (e)
Down (d)	Neutrino (ν_e)
Charm (c)	Muon (μ)
Strange (s)	Mu-Neutrino (ν_μ)
Top (t)	Tauon (τ)
Bottom (b)	Tau-Neutrino (ν_τ)

Why the extra generations of fermions? Why only three? There are some theoretical arguments about why there can't be four or more. Why is the magnitude of charge on the proton and electron exactly the same? What are the rules that govern the masses of these particles? These are questions without solid answers so far.

On the other hand, what did you expect? Science always begins with a question and science must always end with a question. It's never complete. What would be the fun in that?

CHAPTER

EIGHT

FINAL ADVICE

That's it. We have not covered every possible question or combination, but I feel safe to say we have touched on at least 80% of the most important ideas you will meet in your class.

Test Anxiety

My experience is that a lot of students experience some form of test anxiety in technical subjects like physics or math. The experience ranges from slight to paralyzing. Test anxiety is a form of panic, similar to a person drowning. One thing about the person drowning is that they usually make *bad decisions*. That's why it is important to think through these decisions before you are lost in the moment.

The first step is to recognize the beginning signs. This is probably a little different for each person, so notice what happens to you during the test. Is your breath shallower? Take a deep breath. Are you thinking in circles? Stop thinking.

The point is to *take action* rather than let your reflexes control you. When you are losing control, stop and look at the clock. Give yourself 15 seconds. Literally watch the clock tick out 15 seconds (its longer than you think).

Once you have trained yourself to fight back the panic, then we can discuss specific test taking tactics. There are three things to remember.

1. Skim The Test; Do The Easiest First

Usually instructors mix-up the difficultly level, so the easiest ones are not usually at the beginning or end. There is no pattern. You should pick and choose the easiest few and start with those. Getting *something* on paper will boost your confidence.

2. Budget Time; Leave 25% For Review

This is important. It's worth the first couple minutes of the exam. Suppose you have an hour to take a 10 question test. Don't take six minutes per question. This leaves no time for review. Make it five minutes per question, which will leave ten minutes for review.

Make your plan and stick to it. If you have to leave a question partially finished, DO IT! Mark the question with a star and come back to it during your review. You don't want to get bogged down on the hardest question of the test and miss out on the easier points later in the exam.

3. Never Leave A Question Blank

I am always amazed when I get a test with unanswered questions. It ties my hands as the grader because it is impossible to give any kind of partial credit. At least write down the relevant equation. This shows that you are in the ball park. I've even given a few points for the *wrong* equation—at least it shows some effort!

Stay Focused

You and I both know that, in the end, it's your grade that matters. I enjoy learning and I'm sure you do too. But the grade is what goes on the transcript, so think strategically about it.

There are two ways instructors calculate a final grade: using a point system or using weighted average. Mathematically either system can be rewritten as the other, but it does make a difference on how you should look at the work. I like to think about how much each assignment or exam contributes to moving the grade one unit. In other words, if it takes 1000 points to get an A, and 900 points to get a B, then that 100 point exam is worth an entire grade point. The 25 point lab write up is worth 1/4 of a grade point, etc. The same calculation can be done when the grading is a weighted average.

Once you have determined the relative importance of these assignments, you can decide how much effort to put into them. In general, your effort should be proportional to the weight of the assignment. Of course the exams will top the list—you should spend hours studying for them. The thing to avoid is spending hours on something that, in the end, is only worth 2% of a grade point. This also applies to extra credit.

Think of your "return on investment." Your time is your investment. Are you reaping a sufficient return in your grade? If not, move on. You are simply spinning your wheels.

I'll let you in on a dirty little secret. You can't learn all the material. It's impossible and it's designed that way. This is not because your instructor is evil. This is built into the very nature of how school works. The purpose of the system is to expose you to the maximum amount of material. If you are some genius with nothing but time on your hands, maybe you can keep up.[1] But in the real world, you have to make choices.

Don't become too discouraged by a single assignment. One of the things that makes studying physics difficult is the breadth of the subject material. First mechanics, then thermodynamics, then optics, etc. Many of the things you learn in the first few weeks never show up again. One is constantly learning *new* material. The silver lining here is that if you miss something, it may not come up again! So keep plugging away.

---

[1] Even when you think you have learned it, you haven't. You won't really know it until you try to teach it or make something with it.

APPENDIX

GLOSSARY

Acceleration The rate at which velocity changes. Can be a change in magnitude or direction.

Amp SI unit for current, or the amount of flowing charge per second. Typical applications deal with milliamps of current. Symbol = A.

Amplitude The maximum displacement in wave motion.

Angular Velocity The rate at which an object rotates. The speed of its change in orientation. Typically measured in radians per second.

Atomic Mass Unit A non-SI unit of mass useful in microscopic calculations. By definition a carbon-12 atom has a mass of exactly 12 amu. Symbol = u.

Atwood Machine A simple machine composed of one pulley and two masses hanging from either side.

Baryons A three-fold combination of quarks. Both protons and neutrons are examples of baryons. See mesons.

Boson The elementary particles that mediate the fundamental interactions between matter. Every elementary particle is either a fermion or a boson.

Bubble Chamber A contraption filled with supersaturated fluid in which the track of charged particles are easily seen.

Capacitance A measure of the quality of a capacitor. The larger the capacitance, the more charge it holds for a given voltage.

Capacitor An electrical component that stores electric energy. Composed of two conducting plates separated by an isulating material (or vacuum).

Centripetal A force is centripetal if it causes an object to move in uniform circular motion. The word comes from the idea that the force is always pointing toward the center of the circle.

Color A subatomic property of quarks. Completely unrelated to visible color seen with the human eye. Color is the quark-property that is exchanged in the strong nuclear interaction.

Coulomb The SI unit for electric charge. One coulomb is a huge amount of charge: a lightning bolt may carry a few dozen coulombs of charge. It is very typical to deal with microcoulombs of charge. Symbol = C.

Current The rate at which electric charge flows through a wire.

Decibel An alternate way to quantify intensity. Based on a logarithmic scale intended to be roughly correlated with the sensation of sound.

Degenerate Used in the context of atomic orbitals. Two orbitals are degenerate when they have the same energy.

Delta-V A term used in astrodynamics representing the change in velocity required to change orbit. This change in velocity is provided by the propulsion from the rocket engine.

Dependent Variable The quantity that is driven by the independent variable in a mathematical function.

Derivative For a function, the sensitivity with which the output variable depends on the input variable. When the function is plotted on a graph, the derivative represents the slope of the graph.

Differentials An infinitesimally small number. Commonly used by physicists, but mathematically the notion is nonsensical. Replaced with the rigorous notion of a limit.

Diopters The reciprocal of focal length (measured in meters). Used in optometry.

Dipole A system of two equal and opposite charges (electric). Also quantifies the strength of a bar magnet.

Dipole Moment The magnitude of the dipole charge times the distance between them. The far-field magnitude of the dipole is proportional to the dipole moment.

Dynamics The branch of physics that deals with the cause of motion (i.e., force). Frequently contrasted with kinematics.

Elastic Collision A collision in which mechanical energy is conserved. Further, kinetic energy is conserved before and after the interaction.

Electromotive Force The voltage provided by a battery or induced by a changing magnetic field flux.

Electron-Volt A very small non-SI unit of energy. Convinent for working microscopic problems. Defined as the potential energy of an electron in one volt of electic potential. Symbol = eV.

Energy The ability of a system to do work. Energy is either kinetic (due to bulk motion of the system) or potential (due to the configuration of the system). Also see heat.

Entropy Measures the distribution of the internal energy of a system. The larger the entropy, the more evenly is the energy distributed.

Exclusion Principle A statement of the fact that two electrons cannot occupy the same atomic orbital. This is a consequence of the fact that an electron is a fermion.

Farad The SI unit of capacitance. A typical value for a capacitor is in the microfarad range. Symbol = F.

Fermion The elementary particles from which all matter is built. Every elementary particle is either a fermion or a boson.

Field Flux The amount of field that crosses perpedicularly through a certain cross-sectional area.

Field Potential Analogous to the potential energy of a conservative force. Voltage is the common name for electric potential.

Flavor A subatomic property of quarks. Completely unrelated to flavor tasted with the human tongue. Flavor is the quark-property that is exchanged in the weak nuclear interaction.

Focal Point For a lens, the point at which it will focus an image from infinitely far away.

Function A function describes the mathematical relationship between two variables when one of them (the dependent variable) is driven by the value of the other (the independent variable).

Gluon The elementary boson that mediates the strong nuclear force.

Heat A "third" form of energy (as opposed to kinetic and potential). It is now known that this heat energy lost to mechanical systems is really manifest in the kinetic (and potential) energy of its molecules.

Henry The SI unit for self-inductance. Symbol = H.

Ideal Gas A gas is ideal when its molecules do not interact at all. Each intermolecular collision is elastic.

Ideal Spring A spring is ideal when its reactive force is strictly proportional to its displacement.

Image Used in optics. The point in space from which the rays of light from a lens appear to originate. If the light rays actually do run through this point the image is "real", otherwise it is "virtual".

Impedance The magnitude of the reactance of an AC electrical circuit. Analogous to the resistance in a DC circuit.

Impulse Over a specified period of time, the average force times the duration involved. Typically used for collisions.

Independent Variable The quantity that drives the dependent variable in a mathematical function.

Index of Refraction Used in Snell's Law to calculate the refraction angle of light between two tranparent materials. Related to the speed of light in the material.

Induced Current The current induced by changing magnetic field flux. See electromotive force.

Inductor An electrical component that stores magnetic energy. Composed of a coil of wire which may or may not surround a metallic core.

Inelastic Collision A collision which in which mechanical energy is not conserved. A collision is completely inelastic if the objects stick together (no rebound) in the end.

Intensity The amount of energy that flows through a given surface area per second.

Joule The SI unit for energy. Equal to one netwon of force times one meter of length. Symbol = J.

Kelvin The SI unit for temperature. Calibrated to the Celcius scale, but offset so that zero aligns with "absolute zero"—the unachievable state in which the random motion of molecules cease.

Kinematics The branch of physics that deals with measuring motion. Frequently contrasted with dynamics.

Kinetic Friction The type of friction due to the motion of one surface against another.

Latent Heat The rate at which a system absorbs or releases heat when undergoing a phase change. Related to changes in the microscopic potential energy of the system.

Length Contraction A consequence of special relativity. The apparent length of a ruler in motion is shorter than when the same ruler is at rest.

Lever Arm Used to calculate torque. The distance between the axis of rotation and the line defined by the direction and location of the applied force.

Light Rays A mental model which ignores the wave nature of light. Useful when the wavelength of the light is small relative to the distances involved.

Mass Defect The difference between the mass of an atomic nucleus and the mass of its parts. Represents the binding energy released in nuclear fission and fusion via $E = mc^2$.

Mass Spectrometer A contraption using a strong magnetic field which separates molecules by their mass-to-charge ratio.

Matter Waves The proposition that, at the microscopic level, matter does not behave like a point-particle but like a wave.

Mechanical Advantage For a simple machine, the ratio of the output force to the input force.

Mesons A combination of a two quarks. Always involves a quark and an anti-quark. Also see baryons.

Moment of Inertia The measure of the rotational inertia of an rigid object.

Mutual Inductance The magnetic connection between two inductors separated in space. The current in one inductor induces a magnetic field which induces a current in the second inductor.

Natural Frequency The oscillation frequency of a system in equilibrium when slightly disturbed. If an external driving force is tuned to this frequency, resonance occurs.

Newton The SI unit for force. Same property measured by the imperial unit of the pound. Symbol = N.

Ohm The SI unit for electrical resistance. Symbol = Ω.

Orbital The shape of the electron cloud surrounding the atomic nucleus. These shapes govern how atoms and molecules interact chemically.

Parallel Circuit An electric circuit in which the components are connected such that the same voltage is applied across the elements.

Particle A material system of negligible extension. A particle need only be "small" on the length scale of the problem. For example, a planet is considered a particle when analyzing its orbit around the Sun.

Period The amount of time it takes for a system to cycle back to its original state. The reciprocal of frequency.

Permeability A measure of the amount of magnetic force generated within a material by a magnet. A good magnetic core must have high permeability.

Phase The "location" of a system within its oscillation cycle. Quoted as an angle, e.g., a phase of 90° means one-quarter of its cycle is complete.

Phasor A complex number representing an oscillating system. Sometimes presented as a two-dimensional vector because the complex number includes both magnitude (the amplitude of the oscillation) and direction (the phase of the oscillation).

Polarized When the plane of a transverse wave does not change over time. Usually used when speaking of light.

Power The rate at which work is done. A measure of the usefulness of an engine.

Proper Frame A reference frame in which the measurements made are stationary. Used in special relativity.

Proper Length The length measured by a ruler within its proper frame. In other words, the length measured when the ruler is at rest.

Proper Time The duration measured by a clock within its proper frame. In other words, the duration measured when the clock is at rest.

Quarks The elementary particles that form nuclear matter.

Radian A scientific unit used to measure angle. There are 2π radians in 360°. Useful because the arc with length equal to the radius of the circle forms an angle of one radian.

Relativistic Mass A consequence of special relativity. The apparent mass of an object in motion is larger than when the same object is at rest.

Relativity of Simultaneity A consequence of special relativity. Observers in motion relative to one another will *not* agree on which pairs of events are simultaneous.

Resistance The analog of friction in electrical circuits. Ohm's Law relates voltage, current, and resistance.

Resistor An electrical component that obey's Ohm's Law. Introduces power loss is the circuit producing heat.

Rest Energy The relativistic energy that an object has at rest by virtue of its inertia. This is the E in Einstein's famous $E = mc^2$.

Reversible Process An ideal physical process in which no energy is lost when the process is reversed. Frequently used in thermodynamics. A reversible process conserves entropy.

Right Hand Rule Necessary to determine direction in a situation which uses all three dimensions. Used in rotation and electromagnetism.

Self-Inductance The mutual inductance of a single inductor with itself. Is related to the amount of magnetic energy stored by the inductor when operating.

Series Circuit An electric circuit in which the components are connected such that all the current must flow through all the elements.

SI Units International system of units (in French: *Système international d'unités*). Fundamental units include the meter, the second, and the kilogram. Symbols = m, s, and kg, respectively.

Solenoid An inductor used as an electromagnet. Typically includes a magnetic core to increase its strength.

Specific Heat The rate at which a system absorbs or releases heat when changing temperature. Related to changes in the microscopic kinetic energy of the system.

Spring Constant The characteristic property of an ideal spring. The amount of reactive force per unit of length displaced.

Standard Model The quantum mechanical framework that describes how elementary particles interact via the electric and nuclear interactions. Excludes gravity.

Static Friction The type of friction due to the contact between surfaces when they are not moving.

Strain The deformation in an elastic body when it is placed under stress.

Stress The force applied to an elastic body divided by the area upon which it is applied. Stress causes strain.

Temperature The physical sensation associated with heat energy. The average kinetic energy of the molecules in a substance is directly proportional to its temperature.

Thermal Expansion Generally a substance expands when its temperature increases. An important exception is liquid water below 4°C which expands as it gets cooler.

Thermometric Effects Any physical property which changes when the temperature changes. Examples include thermal expansion, pressure, and thermo-electricity.

Time Constant The amount of time required for the property of a system to fall to $1/e$ (about 37%) of its original value. Typically used to describe an RC electric circuit.

Time Dilation A consequence of special relativity. The apparent speed of a clock in motion is slower than when the same clock is at rest.

Torque The force applied to a rigid object multipled by its lever arm. Torque causes rotation.

Transformer An electrical component which utilizes mutual inductance. Usually a pair of solenoids with connected metalic cores.

Turning Points The maximum possible displacement for a particle of a particular energy in a potential energy well. There is no kinetic energy at this point, so the particle must turn around and fall back into the well.

Velocity The rate at which position changes. Captures both the change in the magnitude of displacement and its direction.

Virtual Work The principle that when a system in equilibrium is slightly displaced, the total work involved must be zero. This is true whether the displacement is truly real or merely imagined (i.e., virtual).

Watt The SI unit of power. Equal to one joule of energy per second. Symbol = W.

Wavelength The distance between two adjacent peaks of a wave.

Work The displacement of an object multiplied by the component of force applied in that direction. Doing work on or extracting work from a system changes its total energy.

Work Function The atomic potential energy required to ionize a material. An important property in understanding the photo-electric effect.

INDEX

acceleration, 17
amp, 162
amplitude, 56
angular-velocity, 54
atomic-mass-unit, 229
atwood-machine, 35

baryons, 238
boson, 227
bubble-chamber, 175

capacitance, 161
capacitor, 160
centripetal, 47
color, 237
coulomb, 152
current, 162

decibel, 112
degenerate, 224
delta-v, 83
dependent-variable, 12
derivative, 13
differentials, 13
diopters, 143
dipole, 171

dipole-moment, 158
dynamics, 29

elastic-collision, 78
electromotive-force, 180
electron-volt, 160
energy, 65
entropy, 131
exclusion-principle, 224

farad, 161
fermion, 225
field-flux, 180
field-potential, 158
flavor, 238
focal-point, 139
function, 12

gluon, 237

heat, 118

ideal-gas, 122
ideal-spring, 52
image, 136
impedance, 189

impulse, 83
independent-variable, 12
index-of-refraction, 137
induced-current, 180
inductor, 186
inelastic-collision, 76
intensity, 111

joule, 65

kelvin, 122
kinematics, 29
kinetic-friction, 37

latent-heat, 118
length-contraction, 203
lever-arm, 86
light-rays, 135

mass-defect, 229
mass-spectrometer, 174
matter-waves, 221
mechanical-advantage, 61
mesons, 238
moment-of-inertia, 93
mutual-inductance, 185

natural-frequency, 191

ohm, 163
orbital, 224

parallel-circuit, 166
particle, 61
period, 47
permeability, 169
phase, 113
phasor, 188
polarized, 194

power, 64
proper-frame, 203
proper-length, 204
proper-time, 204

quarks, 236

radian, 8, 55
relativistic-mass, 214
relativity-of-simultaneity, 202
resistance, 163
resistor, 162
rest-energy, 214
reversible-process, 129
right-hand-rule, 172

self-inductance, 186
series-circuit, 166
si-units, 10
solenoid, 179
specific-heat, 118
spring-constant, 52
standard-model, 235
static-friction, 37
strain, 100
stress, 100

temperature, 118
thermal-expansion, 116
thermometric-effects, 116
time-constant, 165
time-dilation, 202
torque, 86
transformer, 185
turning-points, 72

velocity, 17

Index

virtual-work, 63

watt, 65
wavelength, 109
work, 35, 62

www.ingramcontent.com/pod-product-compliance
Lightning Source LLC
Chambersburg PA
CBHW031829170526
45157CB00001B/242